The POWER OF PLANTS

The POWER OF PLANTS

How Nature Shapes Our World

Lucy Summers

The History Press

First published 2026

The History Press
97 St George's Place, Cheltenham,
Gloucestershire, GL50 3QB
www.thehistorypress.co.uk

Cover Illustrated by Lesley Buckingham

British Library Cataloguing in Publication Data.
A catalogue record for this book is available from the British Library.

ISBN 978 1 80399 926 5

Typesetting and origination by The History Press
Printed and bound in Great Britain by TJ Books, Padstow, Cornwall.

MIX
Paper | Supporting responsible forestry
FSC® C013056
www.fsc.org

The History Press proudly supports
Trees for Life
www.treesforlife.org.uk

EU Authorised Representative: Easy Access System Europe
Mustamäe tee 50, 10621 Tallinn, Estonia
gpst.request@easproject.com

There are fairies at the bottom of our garden!
It's not so very, very far away;
You pass the gardener's shed and you just keep straight ahead –
I do so hope they've really come to stay.
There's a little wood, with moss in it and beetles,
And a little stream that quietly runs through;
You wouldn't think they'd dare to come merrymaking there –
Well, they do.

– Rose Fyleman

To my family, friends
(especially Elle, for her unflagging support)
and all those fabulous plants that
make the world a happier place.

CONTENTS

Introduction 11

PART ONE | BEGINNINGS: PLANT EVOLUTION 15

The Day of the Dinosaurs: A World Without Flowers 17
Extinction: The Precipice of a New World 18
Rebirth: The Dawn of Flowering Plants 21
Seeds: Perfectly Packaged Miracles 25
Picky Seeds: Seeds with Needs 29
Pollination: Courtship, Bees and Hoverflies 33
Sexual Reproduction: Sex and Manipulation 37
Soil: Dishing the Dirt 41
Roots: Absorption, Anchorage and Nutrients 46
Alternative Roots: A Banyan Called Brutus 50
Plant Hunters: Curiosities and Commerce 54
The Planet's Lungs: Every Breath You Take 57
Plastic Pollution: Plastic, Not Fantastic 60

PART TWO | SENSES AND DEFENCES 63

Anxiety: Burnout and Stress 65
Chemistry: Chain Reactions and Chemical Warfare 68
Communication: Plant Whispering 71
Gravitation: Defying Gravity 76
Light: Responses to Light 79
Mobility: Plants in Motion 83

Plant Mimicry: Extraordinary Adaptations 88
Sensitivity: Memory and Response 92
Water Diviners: Hydration 95

PART THREE | LIFESTYLES: MODES AND BEHAVIOUR 101

Annuals and Biennials: Life in the Fast Lane 103
Aquatics: Water Babies 107
Bulbs: Buried Treasure 111
Climbers: Soaring Ambitions 114
Desert Dwellers: Cacti and Ephemerals 118
Edibles: Fruit and Vegetables 123
Herbs: Healing and Legacy 125
Houseplants: Plants Out of Place 130
Perennials: Long-Lived Plants 133
Weeds: Chancers and Opportunists 138

PART FOUR | INIMITABLE LIFESTYLES 145

Carnivorous Plants: The Most Marvellous Plants in the World 147
Cycads: Ancient Warriors 150
Palms: Neither Fish Nor Fowl 152
Parasites: Freeloaders and Scroungers 156
Trees vs Shrubs: Why is a Shrub Not a Tree? 161
Trees: Carbon Capture, Congestion and Pollution 164
Foliage: Trees and Leaves 168
Conifers: Evergreen Elders 171

PART FIVE | GAME-CHANGERS: PLANTS THAT SHAPED HUMAN HISTORY 175

Heroes: Contraception, Cannabis and Opium 177
Trade: Cotton, Cocaine and Tobacco 181

Drugs and Narcotics: Poison and Addiction 186
Boozy Botanicals: Intemperance and Sobriety 189
Gentle Refreshment: Coffee, Tea and Sugar 195

PART SIX | NATURE AT WORK AND PLAY 199

Companion Planting: Friends with Benefits 201
Caring: Plant Husbandry 205
Happy Spaces: Flower Failure, Latin and Trouble-Shooting 208
Nurtured by Nature: Stress Less, Live Better 212
Parenting: Backed by Nature 217
Green Fitness: Natural Workouts 221
Walking: A Wonder Drug 224
The Essence of Flowers: Fragrance and Folklore 227
Journey's End: Seasons in Bloom 231
Believe in Tomorrow: Biomimicry and Advances 234

Author's Note 237
Glossary 240
Glossary of Plant Names 243
References 248
About the Author 256

INTRODUCTION

In the most unlikely habitats, across sweeping deserts, deep in tropical forests, in peaty bogs, swamps, woodlands, parks and rivers and even in your own garden, you'll discover daily miracles often unnoticed by the human eye. Genetically programmed to adapt and prevail, plants are rooted in resilience and innovation.

Plants enrich our lives in indescribable ways. I'd venture there's not another natural product on Earth to rival them. Flag-waving from rough grass along roadsides, fields and inner-city scrubland, you'll spy any number of jewelled wildflowers. Vivid corn poppies, misty-clocked dandelions and soft scallops of cranesbill. White fluffs of clover and the midwinter fire of dogwood, every morning sees a new awakening of natural wonders, each superbly adapted to flourish in its local environment.

An orgy of delight and discovery, wrapped in a kaleidoscope of contrasting colour, shape and fragrance, plants might well have been gifted by fairies. Smooth stems, thorny and cork-screwed ones – leaves the size of surfboards and blooms so dainty and diminutive you'll need to get your nose down in the dirt to fully appreciate their beauty. Dazzling and fully biodegradable, it's doubtful any department store, from Saks to Selfridges, has a perfumery to rival floral fragrance. Warm musk, heady notes, zesty citrus or nose-wrinkling whiffs, transporting us down memory lane and subconsciously creating the floral soundtrack to our years.

Whether you subscribe to Earth's creation in the Garden of Eden and that treacherous apple dangling tantalisingly between Adam and Eve in solitary temptation, or endorse the Big Bang theory, I'm sceptical that even *He* could have come up with the intricacies of plant architecture in so little time. Like Adam and Eve, plants wrestle not with conscience but the hardship of daily survival, and it's not always the most talented, best looking or most intelligent that attain success – luck has a great deal to do with it.

Ravaged by goats and on the verge of being wiped out by invasive species, the lovely St Helena ebony, teetering on the edge of extinction, struggled to survive life's challenges. Native to the South Atlantic island of St Helena (where, incidentally, Napoleon Bonaparte was exiled), down to the last two standing and flowering desperately in a last-gasp attempt to set seed while precariously clinging to a cliff face for dear life, she had the good fortune to attract the attention of local guide. The story goes he was so struck by her charms that a makeshift harness was swiftly cobbled together and, scaling the hazardous rock face, he returned with a stem clamped in his teeth. From that single stem multiple plants were propagated and her beautiful white hibiscus-like flowers survived to fight another day.

Curiously, the titan lily (or corpse flower), famously builds up its strength over years to throw out a single massive, stinking bloom. For one week only, this floral Colossus releases a rotting aroma to attract pollinators before dying from sheer exhaustion.

Ever more camera-shy is the century plant, better known as *Agave Americana* – an irony, given that it blooms once in around thirty years as it nears the end of its life, attracting huge crowds to witness one of the rarest spectacles on Earth.

But have you ever stopped to think about what plants actually do? At first glance, trees seem to spend their lives quietly – offering perches and nesting spots to birds, casting shade on hot summer days, then shedding their leaves at autumn's first chill. Tomatoes, from tiny seedlings swell into dense green bushes almost overnight, their stems suddenly studded with fat fruit. Bindweed will grasp and coil with relentless determination, agitating anything in its path.

Have you ever wondered why shade-loving plants enjoy shade? And why is the foliage of some plants grey instead of green? Why do some lack flowers and others perfume? Why bark, why berries, bristles and thorns? Are plants simply harmless pleasures or is there something deeper happening beneath the surface? These and other questions lie at the heart of this book: how plants power the planet and shape our lives.

Crouched in the garden's shade with steepled foxgloves swaying above me, I watch bees slip into their freckled throats and drink the hidden sweetness. Move closer, and you'll see those markings are actually mini-road maps, intricately dotted patterns easily readable by bees that can sense ultraviolet light, guiding them to life-giving nectar. Some claim these beautiful flowers resemble the fingers of gloves and folklore would have us believe foxes wore flowers on their paws when hunting to soften their footfall. Whatever whimsy inspired their name, foxgloves are perfectly crafted to bring us pleasure and vital sustenance to pollinators. But I can't help wondering how these beguiling flowers came to be here in the first place.

Imagine a world where life unfurls with quiet tenacity, each day a testament to survival against the odds. Meet the floral trailblazers, the silent guardians of our planet – plants whose extraordinary talent for endurance and adaptation surpasses anything on a human scale. Through their eyes, you're about to embark on an incredible journey into a realm like no other. Let's return to a primitive landscape when flowering plants didn't exist at all.

Come with me ...

PART ONE

BEGINNINGS: PLANT EVOLUTION

> It is not the most intellectual of the species that survives; it is not the strongest that survives; but the species that survives is the one that is able best to adapt and adjust to the changing environment in which it finds itself.
>
> – Charles Darwin

Over millions of years, plants have evolved in quiet patience, through morphological and physiological adaptations to fixed environments. Often rooted to the same patch of earth, they have adapted, not by wandering but by reshaping themselves – developing traits such as deep root systems for water acquisition, leaf modifications for light capture, adapting physical and chemical defences against herbivores or flowering in sync with pollinators. In contrast, humans, restless and mobile, survive by moving towards opportunity, moulding the world around them, whereas plants thrive by surrendering to a place, relying on genetic changes across generations to meet environmental challenges.

THE DAY OF THE DINOSAURS: A World Without Flowers

A muggy, hot afternoon. It's 33°C (60°F) in the shade, far too hot for hunting under blue skies streaked with wisping cloud. A lumbering Stegosaurus sags in the heat. He has more important matters on his mind – he's looking for a mate. Palaeontologists have discovered that dinosaurs grew fast and died young, and rather like plants, our heat-stressed Stegosaurus is driven by the need to pass on his genes – and quickly. From his bulky height, he takes little notice of the ferns he tramples underfoot.

Drenched in heat, damp air and high humidity, it was only simple ferns and other ancients that flourished. Primitive aquatic plants were uncomplicated folk, typically moss and hornworts, which began colonising the planet around 500 million years ago – around the same time as terrestrial animals emerged. Was it a coincidence that as these plants cooled the planet and boosted oxygen levels land animals emerged? Scientists don't think so. And it's here, when dinosaurs roamed swamps and dense forests, that the story of your garden plants, those freckled foxgloves, perfumed roses, tangled clematis and twining honeysuckle begins ...

Early fossils dating back 470 million years show that the first land plants were similar to liverworts, simple mosses that released spores. Back then, spores contained a single cell, whereas modern seeds have multiple cells. Yet it took another 150 million years for seed-bearing plants to evolve.

Admittedly, prehistoric plants had a limited repertoire, but ferns had a formidable thing going for them – they depended on moisture for reproduction. Releasing spores into the air, which, with light and moisture, grew into small plants known as prothallia, miraculously, they developed male and female sexual organs. Once sexually mature, the male parts released sperm that swam through water in moist habitats, obligingly fertilising the eggs in neighbouring female ferns. Once successful fertilisation occurred, the egg cell grew into a baby fern and a new life began.

Such adaptations to the environment are colossal compared to human evolution. Research shows humans emerged from ape-like status over 6 million years and our human transformation to bipedalism (the ability to walk on two legs) took a further 4 million years to accomplish. A drop in the ocean compared to the feisty ferns' ancestry, which dates back over 300 million years to when their ancient forebears dominated a land bereft of flowering plants.

Such unfathomable timelines demonstrate how plants have struggled to survive, long before humans were a gleam in the proverbial eye, adapting their physical and chemical processes to thrive. And as for our amorous Stegosaurus so caught up in his love-making, how was he to know that his days were numbered?

EXTINCTION:
The Precipice of a New World

In his urgent coupling, our hot-blooded dinosaur was unaware of a devastating asteroid hurtling towards Earth, bringing an end to a dynasty that had reigned supreme for 180 million years. Ah, well, his death must have been swift and at least he died happily. Along with the 75 per cent of all animals and marine life that perished alongside him, the jury's out on exactly what brought the age of the dinosaurs to a close. The asteroid theory argues that a huge blast sent a gargantuan heatwave across the world, soot blocked the sun, impairing plant growth, and as herbivores struggled to survive on depleted greenery, the impact worked its way up the food chain with carnivores eventually starving to death. Others subscribe to the premise that wide-ranging volcanic eruptions across a vast plateau in western India around 65 million years ago sounded the dinosaurs' death knell. Alternative opinion believes the two events converged, precipitating long volcanic winters thought to have lasted for decades, which would have made life extremely tough for our prehistoric friends and played a key role in mass extinction. However.

There's always a *however* in life, isn't there?

The outcome of that job interview you had so desperately set your heart on and two weeks later, you've still heard nothing. The hot date that you thought went so well but they've ghosted you ever since. And budding writers will know all too well the agony of that long-awaited rejection email – *Much as we enjoyed reading your material, I'm afraid we didn't feel ... blah, blah, blah.* Even the inventor of the agile Mosquito, Geoffrey de Havilland, suffered initial rejection from the powers that be when pitching his idea for a new lightweight, speedy plane destined to become the most successful aircraft of the Second World War.

So, it's understandable that ferns struggling to survive in an unstable world inevitably hit a major stumbling block. Spores are slow to mature. Ask any gardener who has ever attempted to gather spores during their fruiting period, and they'll tell you, with a mixture of pride and puzzlement, that it can take months of waiting and watching before knowing if the spores will do their thing and make mini-ferns. And it's not always a reliable method since spores travelling via wind or rain can land just about anywhere. Ideally, they prefer a nice damp nursery, typically found in moss or boggy ground, porous rocks or by ponds and streams, but if spores happen to land in unfavourable habitats, then bang go their chances of reproduction. Incredibly, fossils tell us that ancient species such as royal and cinnamon ferns still thrive today and although many of the earliest ferns are now extinct, modern ferns owe their survival to their dogged ability to adapt in a changing climate.

Incidentally, if you're growing a fabulous maiden hair or gingko tree in your garden, you have a living fossil! Originally from China, before finding its way to Japan and out into the wider world, maiden hair is a symbol of endurance and must be one of the world's ultimate survivors. In the peaceful, calming zen gardens of the Myojoin Temple, Hiroshima, you'll discover one of the most famous veterans of the atomic bomb dropped on Japan during the Second World War. At 150 years old and unchanged for over 280 million years, it's still going strong despite radiation, lightning strikes, storms and typhoons. I'll bet if plants could talk, this courageous survivor could regale us with awe-inspiring tales of derring-do.

Ferns possessed other guiles that enabled their success. While out walking you may have noticed that rabbits and deer avoid bracken (related to ferns), demonstrating another innovative coping strategy. Bracken fronds have a high concentration of toxin that is poisonous to animals (and, incidentally, is carcinogenic in humans) and this stealthy approach assisted their successful leap-frogging from one generation to another relatively untarnished. These unique evolutionary processes kept refining over the millennia but it's enough to know that the origin of today's modern plants – Canterbury bells, crocus, daffodils, daisies, sage, rosemary and thyme – trace their prehistoric lineage to a time when trailblazing ferns battled for survival and fiercely clung on. Today, it's estimated that there are over 10,000 species of fern thriving in very different places, from the Arctic to the tropics, and it's largely down to their battle-hardened ability to react to local conditions and an ever-changing environment that has seen them flourish, both in our private spaces and in the wild.

Life moved on. Climate change and a shift to a drier planet allowed the emergence of a wondrous new empire: the dawn of flowering plants.

REBIRTH:
The Dawn of Flowering Plants

A slimy mass of algae haphazardly washed ashore on a freshwater lake found itself in the right place at the right time, triggering a pivotal moment in history. A fish out of water, if you like, it should have dried up, shrivelled and died. Either by fluke or accident, it survived. Marooned without a drop to drink, incredibly, it became the first plant to settle on dry land.

Fast forward, to the arrival of other seed-bearing plants that rose from the oceans and seas to follow suit. Unlike ferns, these innovative new pilgrims were no longer solely reliant on moisture for reproduction, because flowering plants found themselves in possession of a magical new substance – pollen. A far more effective method of reproduction than water or wind dispersal, plants were suddenly equipped with a previously inconceivable capacity to reproduce through flowers and seed. An altogether more versatile breed, these intriguing floral wonders forced our intrepid ferns into decline, marking the beginning of a beautiful hitherto-unknown friendship. With insects. And bugs were quick to catch on. We have to remember that insects were flightless then, but with plentiful plants growing within easy reach, they simultaneously developed the ability to fly. Unaccustomed to such a rich, convenient food source, they whizzed about this abundant new kingdom, spreading pollen among increasingly promiscuous plants and triggering a population explosion. Competition among plants for the affections of these super-charged flying lotharios was fierce as each began to display individual methods of seduction, developing colour, fragrance, diverse size, leaf shape and structure. Nectar, which came along much later, led to a baby boom, accelerating expansion and giving these awesome newcomers free licence to spread their genetic heritage far and wide.

Nectar is a magical substance; a sugary treat for nectarivores made up of fructose, glucose, sucrose, acids and proteins, making scientists suspect it might play a part in the type of insect individual plants attract. But just making nectar is a mighty risky venture from a plant's point of view. They can use up nearly 37 per cent of their available energy doing so. Which means they need to get the timing right – there's little point producing an abundance of the stuff if you mistime your suitors' visiting hours.

Somewhere along the line – in the Wild West perhaps? – an enterprising cowboy must have thought ... Saddlebags! We need saddlebags! And if, like me, you were wondering if insects were hit with the same inspiration, you'd be absolutely right. When insects developed the remarkable ability to fly, like plants, they also adapted. Better known as coevolution, the physical aspects of both animals and flowers converged, and it wasn't only plants that exhibited new characteristics. Bee fuzz became thicker, they developed leg sacs to carry pollen back to their colonies and their tongues grew longer, enabling them to access the nectar that these new-style plants had begun to produce.

Astonishingly, modern studies show that bees have brand loyalty. Who knew? Bee elders remember plant species where they have previously collected pollen, whereas inexperienced young buckos are less discerning, leading scientists to conclude that bees are strongly influenced by pollen rewards and, astoundingly, are able to retain this knowledge for up to twenty-four hours. Wool-carder bees adore lamb's ear to the point that they will strenuously defend their steeples against incoming marauders. Bellflower blunthorn bees and Scissor bees are decidedly partial to campanula; red-tailed bumble bees have a soft spot for wildflowers and euglossine bees just adore orchids.

Until I began researching this book, I had no idea yellow-faced bees existed, but they do! How delightful. Easily spotted, since they lack stripes and rather resemble flies wearing pale yellow robbers' masks (minus the stencilled swag bags), they are one of the few bee species that, rather than humping all that pollen around, swallow it and regurgitate it for their larvae. With a penchant for garden mignonette, they also thoughtfully waterproof their nests with cellophane-like lining.

Mistakenly, I had naively imagined that all bees were goodly souls. However, as I discovered at a recent beekeeping event, there are good- and bad-tempered bees. Who'd have thought? If you are thinking of keeping a hive, you're advised to buy an even-tempered colony from a local beekeeper, who would be loath to sell you a miserable bunch for fear of besmirching his reputation.

All kinds of things might contribute to their bad temper and even docile colonies can get a bit stroppy when we have thundery weather, although they seem to cheer up again once the weather settles. My teenage nephew, arguably the most easy-going person I know, is known to get a dangerous glint in his eye and turns tetchy when he's hungry. Similarly, bees can get a bit stroppy if there is a nectar shortage – a mood swing most likely provoked when a neighbouring rape field is on the wane or prolonged wet weather prevents them going about their daily pollen rounds. You can hardly blame them. After all, they have families to feed and there's nothing more stressful than an empty larder when everyone is relying on you to bring home the bacon.

Poor handling is another factor – bees appreciate gentle management. It's no good crashing, banging and walloping boxes and frames, it makes them cross. If I were to blunder around your living room, kicking the furniture and swearing loudly, I've no doubt you'd show me the door pretty smartly. Besides, aggression and defence are easily confused and experienced handlers know the difference between a bee with an eternal chip on its shoulder and one that is rightfully defending its corner.

I was also fascinated to learn that the average daily commute from hive to field is normally about a mile, flying at approximately 15–20mph. In case you're in any doubt, bees can travel further afield in their search for food when pushed. A colony of, say, approximately 60,000 apparently thinks nothing of collectively clocking up an impressive daily mileage equivalent to flying to the moon and back. Remarkable people, bees.

Wasps, on the other hand, get a lot of bad press. Less fuzzy and infinitely less loveable, with some species capable of stinging multiple times, and being on the wiry side, they just don't enjoy the same poster appeal. Enjoying a drop of nectar as much as bees and being carnivorous, wasps are less dependent on it and can't carry as much pollen. They are blatant nectar-robbers, quite capable of drilling into nectaries at the base of a plant, drinking their fill and making off without ever coming in contact with a plant's sexual bits. Wasps are ultimate hunters, preying on insects and eating many of the same foods humans enjoy, which explains the hysterical shrieks and panicked flapping at picnics. Rather like trying to evade a particularly persistent admirer at a party with a hyena laugh and an unfortunate emerald gleam of spinach between the teeth, wasps will heckle and agitate until you surrender the goods.

For all the bad publicity, they still do a respectable job in the pollinating stakes. Largely attracted to pale or dull blooms with unusual whiffs, wasps favour shallow flowers with easily accessible or exposed nectar chambers, because their tongues are short. Cow parsley is a firm favourite, and you'll have noticed hoverflies visit it just as frequently. That's because hoverflies are incredible mimics, hoping to dupe the plant into thinking they are true-blue wasps.

As we progress on our magical mystery tour, we'll discover that orchids and wasps have developed an unshakeable fondness for one another, and some are so dependent on these slimline high fliers that some species won't tolerate any other pollinators. But before we meet some fascinating folk from the orchid tribe, let's discover how most flowering plants get their start in life.

SEEDS:
Perfectly Packaged Miracles

From bees to seeds – they don't look very promising, do they? Dull little things really, unlikely to win a bonny baby contest any time soon. But seeds are deserving of an award for the sheer genius of their engineering and are possessed of an inner beauty that is, quite literally, skin deep. I'd argue there's not a gardener on the planet who doesn't experience that curious emotional swell and dampening of the eyeball when they see the first peppering of green pinpricks peeping shyly through the compost.

Given the right conditions, nature's most miraculous transformation takes place. That nondescript seed coat breaks open, a baby root emerges, followed by a special shoot that contains all the ingredients a plant needs to create roots, shoots, spectacular flowers, fruit and perfume.

Inside their all-weather coating, seeds can afford to wait it out until favourable conditions prevail. Excessive dry or wet spells don't bother them, and they can remain viable for years. An Arctic lupin seed still germinated successfully after 10,000 years of dormancy, so you can see this new flowering species was bound to succeed where others had faltered. And while flowering plants might look fragile, you're about to discover they're no slouches in the baby-making department, with individual strategies to secure their lineage.

The plant kingdom is made up of seed-producing and non-seed-producing plants: conifers, which have cones (gymnosperms), and flowering plants (angiosperms), which are seed-bearers. Our old friends ferns, mosses and lichens reproduce by other means, such as spores and fragmentation. Some plants, although not all, have both male (stamens) and female (pistil) sex organs. The stamen houses an anther, a pollen depository, and all it takes for fertilisation to happen is to move the pollen from the male part to the stigma, at the tip of the female pistil. Easier said than done, if you're anchored to the ground with little chance of attracting a suitable admirer.

Fortunately, all those busy insects flying around help plants fulfil their maternal instincts, acting as unwitting couriers by getting heavily dredged with pollen each time they visit a flower. As they fly from plant to plant carrying their precious cargo, pollinators deposit pollen on neighbouring stigmas and, hey presto, conception occurs. It doesn't matter that the receiving plants are a different species, once pollen is dropped, the plant is effectively pregnant and all it has to do is look after its brood until birth. Which happens pretty quickly.

From flower to seed can take roughly two months in many species, and once a plant has reached its objective (and depending on what kind of plant it happens to be), two can things happen: perennials take it easy and just keep up energy production before dying back in autumn and doing it all again the following year, others take a belt-and-braces approach, producing a second flush of modest flowers and, as an added insurance policy, another batch of seed. Whereas annuals and biennials, having sown their wild oats, so to speak, simply die in childbirth.

You already know the basics: that seeds need oxygen, water, warmth and light to sprout. Tough seed coats are all very well, they keep seeds warm and dry, but they exclude the very elements that prevent them from fulfilling their destiny – air and water. What can they do? They are just seeds after all, with no way of breaking out of enforced captivity.

Some plants, such as cucumber and hairy bittercress, have explosive seed pods, pinging their offspring well away from the parent and giving them a better chance at landing in a favourable spot. Some seeds crave light to sprout, others like being hidden in darkness under the soil, while others prefer a light powdering to galvanise them into action, which explains the different sowing instructions on seed packets.

Sweet peas, with their seductive pastels and delightful fragrance, are always a harbinger that summer has officially arrived, but the little blighters have notoriously thick skins, which makes them, alongside beans, beets, cucumber, corn and pumpkin, tough customers to crack. Personal injury or scarring (scarification) is a healthy sadism that many seeds require to get under way, giving them access to the vital air and water they require, so nicking their cases with a sharp knife is the only prompt they need to get their act together. Alternatively, a good overnight soaking will trigger growth just as effectively. When all else fails, they can rely on us to pop them into cosy, warm compost, a gentle watering and exposure to air, and away they go.

In my early twenties and still very much a gardening novice, I decided it was high time to take my pottering to a brave new level and took my first crack at sowing hostas. On the back of the packet were instructions to rub them lightly with sandpaper. It didn't specify for how long or how much. With some intricate surgery, two hours later, I'd produced twenty gleaming white seeds having meticulously sanded off the entire coat of each. I waited with waning hope that they might germinate – you won't be surprised to learn that not one grew. So, human intervention gets seeds part of the way, but what about those that live in the wild and can't depend on our assistance?

Many plants rely on animals and birds to assist in the succession game. Some seeds have clinging burrs that attach to animal fur, thereby hitching a ride to suitable breeding grounds. Others, such as cherries, rely on being eaten so that digestive enzymes and stomach acid break the dormancy. Blackberries and raspberries won't even think about putting their heads above the parapet until they've been foraged by animals or birds. At which point, they are quite happy to strut their stuff and grow into real plants, revelling in unique poop nurseries that initially help boost their growth.

Prairie and chaparral plants have a different agenda. Brittle-leaf manzanita, ceanothus, scrub oaks and wild grasses have heat-resistant casings that can only be broken by a baptism of fire.

With little or no human interaction to help them crack into life, these tough-coated seeds rely on destructive forest fires. Miraculously, seeds can sense rising temperatures, including the chemical signals given off by smoke, and once their super-hard casings are fire-damaged and fires recede, germination occurs. Many evergreen cone-bearing trees are partial to a heat-assisted germination, too, since their armoured woody cones are sealed with resin. Heat melts the highly flammable resin, freeing the captive seeds and allowing them to grow into the trees they've always longed to be. And the unfortunately named Australian snot goblin has gone as far as adapting corky fireproof bark to protect its developing buds from unexpected conflagrations.

You might have noticed the phrase 'coming true from seed' bandied about – which describes whether a seed harvested from a particular plant will grow into a new plant displaying *true* family characteristics. Will the flowers be the same vibrant yellow? Will the leaves look the same? Is it going to be tall or short? You get the idea. That's because seeds harvested in the open might have been pollinated by a different species. Watch your step there. It's always dismaying when, having sown wild seed and expecting a divine pearly pink creation, instead, you end up with an apology of the star plant being sought.

Many seeds are cultivated to improve on their trademark features. Enter the F1 hybrids. These are seeds that have undergone manual tampering to improve on their familial finery and are guaranteed to be Hollywood handsome. Understandably, many of us simply lack the time or inclination to wait for a pecan or conifer to grow, and some plants, I swear, are just damned capricious. You'd need the patience of a saint to grow Himalayan poppies. Their exacting requirements are so demanding as to be exhausting and you're better off sipping a chilled beer in your neighbour's garden and admiring *their* prized blue wonders instead.

Gertrude Jekyll, the famous nineteenth-century British landscaper, cautioned, 'A garden is a grand teacher. It teaches patience and careful watchfulness; it teaches industry and thrift; above all it teaches entire trust.' Since we're not all blessed with patience, and trust is relative depending on where you are on your growing journey, at this point, seasoned gardeners enter the familiar world of grafting, ripe, semi-ripe, softwood and root cuttings – or the realm of the dark arts if you're a novice. Cuttings provide the answer, delivering speed and reliability, where we are rewarded with new plants that are exact clones of their parents within months, without having to undergo years of anxious uncertainty. But typically, one size doesn't fit all.

PICKY SEEDS: Seeds with Needs

Having convinced you that, given the right conditions, all seeds eventually grow into plants, the thought that plants might voluntarily abort their offspring is probably inconceivable, but it happens. Seeds can develop without fertilisation occurring, leaving soft seed traces that lack the trademark crunch of, say, apple or orange pips. (I can't be the only child warned against swallowing apple pips or a miniature tree would grow in my stomach.) Never ones to fail to exploit a lucrative anomaly, fruit breeders seized on this freak of nature to grow seedless grapes and, similarly, have found a way to eradicate those inconvenient glossy, black watermelon seeds. By crossing different watermelon cultivars, they are left with sterile plants with odd-numbered chromosomes that prevent seeds from maturing. Laboratories the world over are exploring new ways to make popular fruits more palatable, although they haven't succeeded yet with apricots, cherries, plums or olives. That's because stoned fruits have a hard tissue surrounding the seed and, to date, no one has managed to produce stoneless fruit. Never say never.

Some seeds are a picky bunch, demanding more exotic rites of passage. Many tree seeds, including lime, maple, hazel, hickory, chokeberry and walnut, are famous for needing exposure to cold, then warmth (stratification), for anything from eight to sixteen weeks before germinating. Nature provides the ideal conditions during wet, cold autumns and winters, which is why autumn-sown seeds do so well, but in hot, dry climates some seeds don't stand a chance. Once again, it's humans to the rescue as we shove bagged seed into the fridge among the groceries, giving them a nudge in the right direction before giving them a swift, luxury blast of heat to warm them up again in readiness to brave the world.

At the other end of the scale, winter-flowering snowdrops face breeding challenges in a season when pollinating insects are few and far between and have adopted a tried-and-tested strategy to produce youngsters. And without ants, snowdrops wouldn't do half as well. Once their pristine blooms fade, snowdrops collapse, and their seed pods containing a special protein-rich attachment – elastiome – grow on the soil's surface. It's an irresistible treat for winter-foraging ants, who diligently drag fallen seed to safe underground nurseries, where it stands a better chance of sprouting without being exposed to above-ground elements and, at the same time, provides a nourishing meal for their larvae. By happy coincidence, seeds remain unharmed, are duly planted and a new generation is assured.

Without ants, there would be no acacias – 'wattles' to those Down Under. And there's nothing a hungry ant enjoys more than the tender, succulent stems of wattle pods. But have you ever sat and watched ants going about the daily grind? With over 12,000 known species recorded globally, ants are just about everywhere: in domestic gardens, high in tree canopies, deep in the soil and barren deserts, isolated outbacks and savannahs. From stretchering the dead to collecting seed, their agility at seed dispersal is legendary.

Ants communicate by laying down chemical scent trails, which accounts for those familiar long lines of military marching, but they also use pheromones to identify and distinguish social status as well as the nest to which any ant belongs. Some species even send out SOS signals by stroking a spike on their hind legs to summon help in emergencies.

A large ant colony lives under my terrace. I can't recall exactly when they set up shop, but evidence of their residence becomes visible each spring as the weather warms up and little heaps of soil suddenly litter cracks in the paving. It's embarrassing to admit how many hours I've sat transfixed by their industrious activity, and I make no apology for sitting doing nothing. Watch them long enough and you'll realise that ants are an outstanding tribe. As the years pass the colony has extended hugely. The small village has now expanded into a sprawling subterranean metropolis under my feet. Thousands of ants (maybe millions, anyhow, far too many to count), all working together to create a hugely sophisticated community.

Today, some of them are dismembering a beetle. Yesterday, it was a moth. The day before, a grasshopper dissecting corpses with the precision of a Formula One pit crew. I was dumbfounded to find an explanation of this intricate surgery that read like an episode from a medical drama. Astonishingly, carpenter ants carry out life-saving surgery on their fellow nest mates. Research from the University of Lausanne discovered that not only could they detect the location of injuries, but to some extent can diagnose and treat wounds to maximise the wounded's chance of survival. Ants suffering with femur or thigh injuries received pre-amputation treatment, with ants licking the wound before removing the damaged limb. Yet injuries sustained in the lower leg were subject to the same licking but no amputations were carried out. I knew it! Each of my ants seem to know their role: hunters, gatherers, medics and lookouts, while others form an orderly line and tramp back home to a large acacia where visible holes as large as thimbles at its base are their points of entry and exit. It's always staggering to witness the weights they carry. It has been long established that ants can easily heft in excess of 100 times their own bodyweight, which is proven as they hoist leaves and bug debris with consummate ease but quite marvellously;

recent studies show that the common American field ant can cope with pressures 5,000 times heavier.

As importantly, these extraordinary insects form an vital part of the food chain and are champion recyclers, playing an indispensable role in waste management and redistribution of nutrients. And not just in our gardens. Recycling and aerating the soil through their intense tunnelling, studies show that ants are responsible for removing over 50 per cent of available food resources from rainforest floors, leaving the remaining debris for other animals to clear up. Their social structure forms a valuable framework, displaying a fascinating division of labour that influences resources acquisition and, consequently, they are unique lynchpins in the environment. Oh, to be an entomologist!

From ants to jackfruit, a tree that requires more hefty helpmates in its bid for successors. Jackfruit happens to be a bit of a delicacy if you're a macaque monkey. Relying on a diet rich in fruit and plants, long-tailed macaques and the largest tree fruit in the world make a harmonious duo. Indigenous to South-East Asia, macaques have received a lot of bad press in recent years, being forced from wet lowlands and dry forests through habitat loss to interact with humans as never before. Blessed with superior intelligence, you'll find endless online videos of these cheeky opportunists stealing tourists' sunglasses and bartering them back for a snack.

Typically, jackfruit bear hefty yellow, egg-shaped fruits some 2ft long and weighing roughly 10kg, which ripen in the rainy season. Clever as they are, macaques can't tell a ripe fruit from an unripe one without assistance. Enter the giant Malabar squirrel, who, with its acute sense of smell, can sniff out ripe fruit in a flash. The cunning macaques wait until a squirrel has identified a luscious fruit and then shamelessly mug him of his find. Using sharp teeth to rip away the tough skin to get to the tender flesh inside, a game of *I got it first!*, or in British parlance, *Bugger off!* ensues, with the squirrels valiantly trying to retain their prize while being constantly batted away by the monkeys. If the battle is lost and with the fruit still too large to handle, macaques gnaw through the stems, sending fruit smashing to the ground and then by digesting, defecating or spitting out seeds, they give jackfruit offspring their start in life.

As impressive is their use of tools. Macaques, chimpanzees and capuchin monkeys use stones, rocks and sticks in daily tasks for cracking nuts or fishing for termites. Macaques have even been seen washing sand off fruit before eating it and while birds may not be as dextrous, Caledonian crows make use of sticks to probe for food and black parrots use stones to pulverise shells before eating the crushed powder and offering the regurgitated snack to woo females.

And plants are no less inventive in their quest for seduction ...

POLLINATION:
Courtship, Bees and Hoverflies

Bees get all the credit for pollination and are responsible for producing most of the food we eat and pollinating roughly 80 per cent of all flowering plants worldwide. Without their benevolent industry, much of the fruit, vegetables and other sweet treats like honey, wine and chocolate would disappear from our supermarkets. Gardeners are encouraged to grow nectar-rich plants that attract pollinators for good reason. It helps maintain a healthy garden balance and provides essential food for valuable insect populations, but most people seem to think it's just bees and butterflies that do all the heavy lifting. Bumble bees, alongside cats and dogs and teeny-weeny bunny rabbits, prove certain species are deeply ingrained in human affection, so how is it that hoverflies are so overlooked? I bet you've never given them a second thought.

Not nearly as stripy or fuzzy as bees, hoverflies have an image problem. Highly migratory, trillions of high-flying hoverflies make seasonal pilgrimages each spring to southern Britain, rarely receiving the accolades accorded to buzzy bumbles, although they are thought to be the world's second most-prolific insect pollinators, visiting over 70 per cent of global crops. And what a fascinating tribe they turn out to be.

So convincingly do they imitate the behaviour of bees and wasps that they are often mistaken for them. Master mimics with no capacity to sting, they're more likely to lick your arm with their long tongues than deliver a stinging reprimand. Admittedly, they are unable to carry as much pollen as bees, but being superb aviators, they fly greater distances, travelling hundreds of miles a day, distributing billions of pollen grains over both land and sea between Britain and Europe.

It's impressive to learn that hoverflies eat 6 trillion aphids and visit billions of flowers seasonally, and just when I thought my admiration for their industry could soar no higher, I learned they are not beyond laying their eggs in bees' nests, where you'd think they might receive a hostile welcome. On the contrary. Instead of disrupting a hive's harmony, bees gladly accommodate hoverfly larvae because hatched youngsters offer an efficient cleaning service by conveniently hoovering up undesirable debris.

So busy at playing good Samaritans, hoverflies are seemingly unaware of their market value, although some plants use cunning and deceit to attract these super-charged pollinators. Orchids can be a capricious crowd with exacting demands. Native to south-west China, the mysterious slipper orchid proves to be a practised deceiver. Its bronzed, mottled flowers display white tufts that, from an aerial view, might easily be mistaken for an aphid colony by any self-respecting hoverfly. To make a convincing show of things, the orchid adds a dollop of honeydew while simultaneously wafting a fake aphid scent, hoping to lure them with the promise of a hearty meal.

All potential parents, whether animal, human, reptile or insect, seek hospitable housing when they are thinking of starting a family, and our hardworking Mrs Hover makes a beeline (pardon the contradiction) for the orchid in the belief she has found decent lodgings to raise her kids. Impervious to the con, mum lays her eggs, getting dredged with pollen in the process, and the wily orchid, now royally fettled, can sigh in relief that its ruse paid off.

No less calculating is the Christmas star orchid with gleaming white, waxen petals and distinctive spurs (extra-long nectar tubes), which is equally exacting in its choice of bedfellows. There is only one romancer it's hoping to attract – the African hawkmoth, a pollinator with a tongue long enough to do the job. But at least the Christmas star has the good grace to offer post-coital rewards. The British bee orchid defies family tradition, employing more self-reliant traits – a self-pollinator, she simply drops pollen on her own female parts to fertilise her flowers.

Some plants are highly inventive in their bid for fertilisation, and not beyond harnessing sex slaves to suit their purpose. To look at an elegant evergreen bull bay, or magnolia, with its opulent waxy chalices wafting lemon-vanilla fragrance, you would never imagine this aristocratic shrub has ever practised deception in its life. More gaoler than breezy first date, its seductive cunning is driven by necessity when, millions of years ago, it heavily relied on beetles for pollination. An intriguing behaviour so ingrained that modern-day magnolias still practise the same crafty ruse today. Closing off the female stigma, it traps beetles overnight inside the flowers, liberally dowsing them with fresh pollen. Come daybreak, the love slaves are released to fly off and pollinate receptive blooms.

And some plants don't bother investing valuable resources such as nectar, showy flowers or fragrance to attract pollinators. Many of the world's most important crops are wind pollinated, including barley, corn, rice and wheat. So too, are firs, pines, spruce, hazelnuts, pecans and walnuts, which produce larger quantities of light, dry pollen from small, plain flowers that is carried on the wind. Not being an exact science, you can imagine that the majority of pollen goes to waste but enough lands on target to justify the process.

Unpleasant body odours do little for human courtship but, as I discovered years ago, a nasty niff can be just the thing a plant needs to attract attention. Shortly after moving house, a strange plant put in an appearance. I watched its swift growth with interest on account of the fact I had no idea what it was. Tightly wound umbrellas rapidly unfurled into fabulous marbled foliage, followed by spectacular deep-burgundy, lily-like spathes. A few days later, a polished purple-black appendage emerged at its centre. From a design point of view, it was a stunner.

A few minutes thumbing through my plant encyclopaedia revealed it to be the exotically named voodoo or stink lily. Or to be politically correct, *Dracunculus vulgaris*. Deserving of a starring role in any sci-fi film, its moody purple blooms (which aren't flowers at all but bracts), are truly eye-catching and the vulgar tag comes from its scent. Numerous red and purple flowers have this peculiar reek, not exactly the stench of rotting flesh but certainly an unpleasant boggy whiff that barely lasts a few days or so. Certainly not off-putting enough to make me reconsider giving my voodoo lily garden room – its handsome swagger far overshadows its malodorous personality. The foul smell attracts its own admirers by way of beetles and flies – proving some people just love a bad boy! And what a stealthy seducer it turns out to be. That astonishing (if you'll pardon my indelicacy) phallus sits over a lumpy chamber that houses both male and female flowers. Attracted by the stink, flies slide down the smooth, shiny spathes and are then held hostage for twenty-four hours. The slick surface prevents them from clambering out again and as the captives stumble about trying to find an escape route, they dust pollen over the stigmas of female flowers. At which point the male flowers burst, their spathes wither and the exhausted sex slaves are finally liberated.

Bonneted columbines are famously promiscuous, and it won't matter if you've elected to grow the enchanting, spurred *A.* 'Blue

Butterflies' or a crowd of gorgeous spidery *A.* 'Yellow Queen', because those mischievous pollinators will sprinkle pollen willy-nilly across your aquilegias and over time, you'll find your restrained colour chart is all over the place.

In the plant world, the dating game is often no more straightforward than in ours, and some plants, including many apple and pear cultivars, are just plain sterile, lacking the ability to produce homogenised offspring without some sort of intervention. For the record, if you're growing apples, any crab apple will serve the purpose – they are more than happy to service all comers belonging to the apple family.

Other plants are simply unable to mirror identical family features, owing to their method of reproduction, which might be a blessing in disguise for some. The stinkhorn mushroom, with a face only a mother could love, reproduces by spores, followed swiftly by my pet hate, the sludge-coloured *Rosa* 'Viridiflora', which is resoundingly sterile. Surprisingly, infertility is no barrier to breeding. Plants didn't get to where they are today without getting artful and they use all kinds of emissaries from breezes to bees to get the job done.

SEXUAL REPRODUCTION:
Sex and Manipulation

The media would have us believe that men think about sex every seven seconds. Although it's a good headline, if it was true, I doubt they'd get anything done. In reality, research shows that men think about sex about nineteen times a day, and women, half that number. Interestingly, men thought almost as much about food and sleep, making me wonder (although this a speculative leap lacking any scientific evidence) whether men still harbour latent hunter-gatherer instincts? Anyway, if we could read plants better, we might find their preoccupation with sex is significantly higher – and the assumption that all plants are hermaphrodites is equally flawed.

Charles Darwin (1809–82), an English naturalist whose theory of natural selection provided the foundation of modern evolutionary studies, revolutionised our understanding of the natural world with ideas that were inconceivable at the time. Together with other botanically inclined thinkers, he provided profound insights into the incredible diversity of life on Earth – the origin of both the human species and plants.

As we've seen, numerous plants possess both male and female reproductive organs, negating the need to mate, since they can do the job quite neatly alone. So, it's surprising they do everything in their power to avoid self-sexing for fear of the damaging effects of in-breeding. Darwin was among the first botanists to ponder the contradiction, baffled as to why plants should ever need to have separate sexes if they were capable of doing it themselves. And he concluded that the most likely explanation was that hermaphrodites would possibly benefit by becoming experts in one sexual function, rather than being Jacks of all trades. Put simply, if a plant is going to divert all its effort into making leaves instead of pollen, seed production would fly out of the window, limiting the opportunity for expansion.

Nothing sells like sex. Films, books, tabloids and kiss-and-tell scandals keep the world enthralled. Plants, of course, don't crave erotica – but they are relentless seducers, flaunting colour and perfume and flirting shamelessly with bees, birds and breezes for one end: reproduction. Fortunately, plants don't go in for complicated emotional relationships – no need to find a soulmate or argue about who does the housework or childminding. They don't even need to be in the same room, because plant reproduction can happen in diverse environments – no hand-holding or romance required. Such economy ensures a parent's genetic characteristics are passed on, allowing plants to produce kids in a relatively short time compared to humankind. You have to remember that plants have no emotional attachment to their offspring, and to ensure their successors, like most living things, they procreate either by sexual or asexual reproduction. At the risk of stating the obvious, let's briefly define the difference.

Sexual reproduction requires two parents to contribute genetic material to produce unique kids. Some plants have both male and female cells (gametes) and through fertilisation, the combined DNA from two separate plants or parents produces seed – or children, if you like. The genetic material donated by both parents means that their offspring aren't identical to each other, giving rise to diversity which can help them adapt to environmental changes.

On the other hand, asexual reproduction allows a single parent to produce offspring that are identical to each other and exact clones of the single parent. Both methods have their pros and cons, which is why some organisms do both to maximise their chances. While the asexual process is incredibly efficient, inevitably, there are drawbacks. As well as passing on positive influences, offspring also inherit family flaws, with kiddies liable to fall prey to the same deficiencies and diseases as their parent. Since the kids all descend from a single organism, they are clones of each other, precluding the variety of diverse physical and emotional characteristics seen in human procreation.

And of course, should there be a disastrous environmental event, it can wipe out an entire species, evidenced by Dutch elm disease, first identified in north-west Europe in 1910, which rapidly spread to the UK and Scotland, wiping out between 10 and 40 per cent of elms. Fortunately, by the 1940s, it had died down and scientists at the time predicted that unless the fungus altered significantly, another epidemic was unlikely. Like COVID, they weren't expecting a second, infinitely more aggressive wave caused by an entirely different species. But that's what happened. In the late 1960s, a fatal fungus spread by elm bark beetles on logs imported from Canada to the UK wiped out these elegant trees in their millions.

And what about family traits? Did you inherit those flapping ears from your father or that appealing pert nose from your mother? Are those gorgeous, sculpted cheekbones from granny? Or do your parents swear blind you must have inherited that unruly red mop from your great, great, great-grandfather twice removed. Ah, the theory of genetics, attributable to – would you believe – sweet peas.

Although perennial sweet peas lack the gorgeous colourways and perfume of the seductive annuals, they can be relied upon to flower year after year with little intervention. (By the way, if you want to keep sweet peas blooming for as long as possible, pick the flowers daily or they'll revert to survival mode and quickly make seed pods.) However, a rather more significant discovery by an Augustinian monk, Gregor Mendel, in the late eighteenth century proved revelatory. You have to hand it to the thinkers, philosophers and botanists of the period, challenging known norms and discovering new ideas and technologies that proved to be world changing. Our stoic monk, studying mathematics and sciences at the University of Vienna (twice failing to gain his teaching certificate), was greatly distracted by inherited genetics.

Yet, academic failure didn't deter him from conducting what must have been a tireless study, which overrode all perceived wisdom of the time – that children's physical characteristics were a diluted blend of their parents. In what must have been an excruciatingly laborious process, from 1857, he grew over 10,000 ordinary sweet peas, meticulously detailing each plant's height, pod shape and colour. Being hermaphrodites, peas possess both male and female cells, which means they can self-fertilise. Mendel observed that when peas were left to their own devices and fertilised themselves, certain physical characteristics were consistent in their offspring. In other words, had he been breeding horses and not sweet peas, the foals would strongly resemble the parents. Dominant genes, he concluded, would always be displayed in a first generation and recessive traits popped up in the next generation, just as long as plants self-fertilised.

Family traits are all very well, but no self-respecting parent can think about raising a brood without first putting down roots. Unsurprisingly, growing mediums and soil are the most important elements in a plant's development. And as it turns out, soil is a very interesting thing indeed.

SOIL: Dishing the Dirt

As caviar is to lumpfish, so soil is to dirt – and it wasn't until I got into horticulture that I came to realise that soil is a pretty marvellous thing. Without a decent drop of the stuff, plants are beggared before they even begin. Would you believe that just a quarter of a teaspoon of soil contains up to 10 billion living organisms, outstripping the number of people on Earth? To date, a mere 1 per cent of soil microorganisms have been identified and with such vast microbial diversity, it's not surprising we still know so little about it.

Soil is what lives under the dirt. A living, breathing organism that continually absorbs oxygen and releases carbon dioxide. A complex, earthy layer packed with bacteria, minerals and organic matter from decomposing plants plus vital air and water. Soil provides living quarters for millions of beneficial creatures and bacteria, and all terrestrial life depends on it staying healthy.

Earthworms are super-efficient recycling centres. As they tunnel through the darkness, their slow, relentless excavation simultaneously blends, oxygenates and loosens soil as well as creating advantageous drainage channels to filter excess ground water. Eating everything from fallen leaves to dead animals, plants and fungi, their signature worm castings contain an impressive 1,000 times more beneficial bacteria than the original soil they ingest – and five times more nitrogen and phosphorous.

Dirt, on the other hand, is what you're left with when all these magical ingredients are depleted. In reality, most of us get the soil that comes with a house purchase or rental, and if you've bought a house on a new development, you can bet your bottom dollar that the developer has left nothing behind worth having. Universally, builders scrape off all the good stuff to dig house foundations and what impenetrable lumpiness is left behind isn't worth a light.

Medieval farmers had a workaround for unpromising soil and commonly travelled for miles to load up their carts with chalk and lime to enhance indigenous heavy clay to boost crop productivity in a process known as marling. They would till the marl together with well-rotted manure in their fields, transforming stubborn clay that is notorious for compaction, which impairs drainage and air retention, and transforming it into a fertile, workable tilth. If you look at it from a plant's point of view, it's so much easier getting young roots down into an expandable substance that allows free growth and a wealth of nutrients. I've always imagined it must be similar to kicking off a tight pair of shoes after a working day and instantly experiencing that rush of relief as the blood flows back to your feet.

Aside from providing for plant life, soil helps mitigate climatic events, soaking up vast quantities of water during heavy rainfall or flooding and minimising the effects of drought by drawing on water reserves during prolonged dry periods. As well as filtering out pollutants that pass through its sponge-like layers, it's also a gigantic carbon store. Second to the oceans, soil is the planet's second-largest active pool of carbon. However, the soil's capacity to retain vast amounts of stored carbon is compromised by the continuous exchange of carbon between soil, the atmosphere and plants – driven by unsustainable land-management practices and changes in land use.

Most of us are acutely aware of carbon dioxide, a major greenhouse gas that traps heat in the atmosphere and warms the planet. It is largely produced by wildfires, the burning of fossil fuels, coal extraction, natural gas and oil production, as well as by natural events such as volcanic eruptions. According to NASA, human activity has increased atmospheric carbon dioxide levels by 50 per cent in less than 200 years. This highlights why soils are so vital in the climate crisis as they store more carbon than the world's atmosphere, plants and forests combined.

Soil is also nature's larder, giving life to 95 per cent of the crops in our fields, trees in forests and the flowers in our gardens. But we are losing topsoil at a worrying rate, somewhere between ten and forty times faster than it is formed. Just as our bodies repeatedly replenish our skin by shedding old cells and creating new ones, new soil is formed daily – but boy, oh boy, is it a slow process. According to the Soil Society of America, it can take up to 100 years to produce just an inch of topsoil in warm, humid climates, and a sobering 1,000 years in colder, drier areas. That puts soil health right up there on the eco-agenda.

As new soil is formed, so it can die, and humans are pretty good at killing it. Excessive agricultural practices, the use of chemical fertilisers and pesticides, coupled with excessive cultivation, might mean cheaper food in the short term but in the long run, it's incredibly damaging to delicate soil ecosystems. But why should it be an issue? We don't have to look far back in history to find out.

The Great Dust Bowl crisis of the 1930s affected the south-west Great Plains of America when the enacted Homestead Act of 1862 promised an irresistible opportunity. For a nominal fee, any US citizen over the age of 21 (as long as they had never borne arms against the government) could claim 160 acres of allocated land. A once-in-a-lifetime opportunity that drew soldiers, farmers and African-American labourers to take up the challenge. I'm sure that President Abraham Lincoln was well intentioned when he made a speech at New Haven, Connecticut, in 1860, espousing his hopes for a better future:

> I don't believe in a law to prevent a man from getting rich; it would do more harm than good. So, while we do not propose any war upon capital, we do wish to allow the humblest man an equal chance to get rich with everybody else.

It's doubtful he could have envisioned the disaster that followed, some seventy years later.

Lacking basic tools or the financial means to acquire farming equipment didn't dissuade impoverished settlers reaching for a slice of the American dream. Motivated by the promise of land ownership, they faced severe hardship and environmental challenges, and while the Act required settlers to implement only minor land improvements over a five-year tenancy, its vague provisions often worked against them. As a result, much of the land ended up in the hands of speculators, cattle ranchers, loggers, mining companies and railroads. In fact, of the 500 million acres originally offered, only 80 million acres were claimed by genuine homesteaders. As these stalwart pioneers ploughed up hundreds of acres of traditional grasslands, replacing them with wheat or cattle ranching, regions became overgrazed and overcropped. Demand for wheat grew, prompting thousands more acres to be turned over for increased grain production and resulting in the destruction of traditional prairie lands.

Eventually, unable to sustain its new tenants any longer, the unthinkable happened. Coupled with powerful winds, the arrival of severe drought in the 1930s turned the land to dust – and it simply blew away. Great dust clouds mushroomed across the skies, burying homesteads. Grit was everywhere as once moderately fertile land was stripped bare, leaving behind a lifeless dust bowl. With their dreams in ruins and unable to wrest a living from the soil, the settlers abandoned their claims and were forced instead to eke out an existence as migrant workers. Best known for his Gaia proposal in the 1970s, British chemist James Lovelock lamented, 'Sadly, it's much easier to create a desert than a forest.'

FROM WASTE TO SPACE

With a lifestyle as endearing as their name suggests and measuring less than 1mm in length, water bears, also known as moss piglets or tardigrades, thrive in an astonishing range of environments: the Arctic, deserts, forests, grasslands, ponds, lakes and seas. Renowned for their extraordinary resilience, these stout little creatures have survived sub-zero temperatures, intense radiation and, astoundingly – space.

Predating the dinosaurs, tardigrades are among the earliest species known to man and often the first organisms to colonise new habitats, sustaining themselves on plants while serving in one of the world's most vital professions: recycling. By consuming plant debris and decaying organisms, they return nutrients to the soil at a phenomenal rate and, at the same time, occupy a crucial place in the food chain while doubling up as environmental barometers. Their absence often signals pollutants, while abundant populations point to high-quality air and water. Almost indestructible, these microorganisms are probably beneath your feet right now – capable of surviving wild temperature swings of up to 150°C at either extreme and, astonishingly, they can live up to thirty years without water. When threatened by dehydration, they simply retract their head and legs, expel nearly all body water, curl into a ball and play dead – entering a dormant state that ends the moment they sense moisture. Such incredible survival abilities fascinated scientists. How can such tiny creatures withstand extremes that would obliterate most life forms?

Perhaps their greatest feat is one of resurrection and recolonisation. In June 2005, NASA launched 3,000 water bears into orbit – and an astonishing 68 per cent survived. If they can withstand a hostile environment guaranteed to kill any other organism, could they play a role in colonising extreme or degraded habitats? Beyond their ecological importance, scientists are investigating their potential in fields ranging from testing atmospheric conditions in space to improving human health. Could tardigrades unlock solutions for long-term vaccine cold storage? Might they help protect cells from chemotherapy-related damage or radiation exposure? Small as they are, their remarkable versatility continues to inspire innovations once thought impossible, reminding us of nature's boundless ingenuity in addressing humanity's greatest challenges.

Of course, no organism, however talented, can think about raising a family or forming relationships without putting down roots, and here we discover a mysterious underworld chiming with purpose ...

ROOTS: Absorption, Anchorage and Nutrients

Rooted to the spot, it's impossible for plants to move to a more desirable neighbourhood, seek shelter in a storm or flee when danger threatens. And being anchored to the ground makes it damned nigh impossible to go looking for a mate. But they meet these challenges with infinite determination. Largely hidden underground, nobody takes too much notice of roots. Manifesting in all shapes and sizes, and despite adversity, hostile terrain, weather and life-threatening challenges (including the all-too-common black-thumbed gardener), you might develop a whole new healthy respect for these grubby, unsung heroes when you realise the colossal responsibilities they bear.

Although lacking the good looks of their above-ground attachments, roots are nothing short of amazing. Strong enough to crack rocks and equipped with sensitive tips to absorb water, dissolved minerals and other valuable nutrients. Whether roots are aerial, clinging, climbing, stilted, buttressed or tapped, they all share a common purpose – anchorage. We need only look at the architecture of ancient castles, standing proud and impenetrable, their substantial walls and buttresses offering greater strength and stability, to wonder if, many centuries ago, a master builder studied a sturdy oak and came up with the idea of defensive architecture.

After anchorage, the roots' main function is sucking up water and nutrients, and in some cases, absorbing oxygen from the air or fixing nitrogen. Focused on those fat, dirt-crusted, home-harvested carrots, you might barely have noticed that they bristle with a multitude of fine, branching hairs perfectly designed for water absorption. Beetroot, carrots, ginger, parsnips and yams have tap roots, a single thick main root, branching into a thicket of secondary and third-level roots and culminating in rootlets at the tips that play a vital role in water absorption. Tap-rooters are deep-soil divers, establishing tenacious foundations. A simple but effective survival tactic that provides firm footing, allowing plants to absorb all the goodies they require. Those of you who have tried pulling up dandelions or milkweed know just how difficult it is to dislodge the blighters, and we're usually forced to resort to a spade or weed torch to get them to surrender their home bases.

Aside from their obvious function in anchoring plants in place, roots are the engine rooms that keep all the visible parts of a plant motoring, as well as doubling up as underground explorers. While the main root works as a stabiliser, plants rely heavily on the assistance of those secondary and tertiary roots lying nearer the surface to extend their absorption capabilities. Varying in size and appearance from conical carrots to swollen roots, such as radishes or thick, fleshy, tuberous potatoes, on the face of it, it's a simple enough survival mechanism. However, the magic lies in the hydrostatic pressure that roots exert to force fluids upwards to the upper parts of the plant, making them the most important part of any plant. It's a very good reason why roots hide underground, making them less susceptible to damage or being eaten by foraging animals. Newbies to seed sowing should take heed: a plant may grow new leaves but damage a juvenile stem or root and they are incapable of growing replacement parts.

Talking of nitrogen, it's an essential for plant growth. It's what makes our plants and lawns lush and green. Veggie gardeners soon learn that beans are a handy crop with a surprising ability to take nitrogen from the air and trap it in the soil. Great news! Surely that means there's no need to add fertiliser when growing your own produce? My apologies – things are rarely as straightforward as they seem. While knobbly lumps on their roots means that legumes and members of the bean clan have the ability to fix nitrogen in the soil, the benefits disappear as soon as they're harvested because nitrogen peaks during flowering. After which, being savvy hoarders, plants move this essential gas to their seeds where it is converted into protein building blocks. But, if you insist on cutting your bean and pea plants, you're removing a powerful growth-booster.

Annoying, isn't it? Just when you thought you'd found a shortcut to humping around heavy barrows of organic matter. Never mind. If you've implemented crop rotation on the veg patch, there's a nifty nitrogen trick that is immensely useful in enriching soil for future crops and also helps eliminate soil-borne diseases. A surefire method of hanging on to that valuable nitrogen for future benefit is growing green manures such as rye and vetch. Usually sown in autumn, allowing time for roots to develop before winter downpours wash soil away, green manures do a sterling job of binding and feeding light soils and opening up heavy clay to help improve its lumpy structure. Growing happily away over winter and into spring, green manures are then turned and dug back into the soil, neatly preserving essential nitrogen. Sowing catch crops such as radish and mustard between your star crops is another great way of returning nitrogen to the soil and, thankfully, helps suppress weed growth in the process.

You'll have noticed most garden perennials have a tangle of threadlike strands – fibrous roots. With no single, stubborn main root like carrots or parsnips, fibrous roots operate in the same manner, focused on support, respiration and absorbing vital minerals and water. Without needing to store food in the same way as root vegetables, all their energy goes into making shoots, leaves, flowers and fruit. Bamboo, barley, elm, oak and pine, lilies, geraniums, grasses, palms and rice, among countless others, all depend on fibrous roots for survival but have no need to anchor themselves as deeply. Their roots still manage to achieve a good fixing depth by spreading horizontally, allowing them to cover a wider area, where they can maximise the intake of oxygen, water and essential nutrients. Some roots, such as mangroves and corn, defy the norm, preferring to grow above *and* below the soil, or even in the air. Both have fibrous roots but are braced with propped roots above the soil that angle downward, maximising moisture and nutrient absorption, plus those extra guy-lines help them withstand any gusting winds.

The ability of roots to explore and mine resources from the soil helps plants to adapt and withstand various environmental stresses, such as drought or nutrient deficiencies. Amazingly, roots can detect physical changes and respond appropriately by bracing for impact by thickening their cell walls.

Aside from their obvious function of keeping plants standing upright, storage and sending food to leaves and flowers, roots guard against soil erosion – an absolute necessity if they want to retain a ready supply of food. If soil is washed away, plants are in danger of losing their footing or dying from starvation.

Despite being hidden underground, roots are a sociable crowd and readily form beneficial relationships through mycorrhizal fungi growing in association with other plants. It's a symbiotic friendship whereby major soil nutrients that underpin cell development and division are exchanged. These complementary fungi extend plant root systems by increasing their food and water intake, as well as boosting drought tolerance.

Sadly, not all fungi are well meaning. Watch out for honey fungus, a fungal disease that, despite its appealing moniker, is a common and potentially fatal disease among shrubs and trees. In private gardens, despite there being no competing fungi to keep them in check, you won't have to worry about it much, but in open land, especially that populated with old tree stumps, you can come severely unstuck. The galling thing is you probably won't notice plants are affected until it's too late. Symptoms are slow to appear and can be attributed to any number of other causes – branches dying back, flower failure or unusual cracks in bark – and by the time the tell-tale tawny-layered mushrooms (that may or may not pop up in late summer or autumn) appear it's too late. Unobserved underground, invidious white webs accompanied by sneaky black bootlaces (rhizomorphs) are relentless in their spread, eating away at roots and eventually causing the whole plant to collapse. No miracle cures either, I'm afraid. Naturally, perennials are more easily replaced but large shrubs and trees prove greater losses. The only way out is removing the entire thing, roots, soil and all, thereby cutting off the fungi's food supply.

However, some plants are rooted in menace from the day they were born ...

ALTERNATIVE ROOTS: A Banyan Called Brutus

Brutus had an unfortunate start in life. Abandoned by his parents as a seed and subsequently dumped by a careless bird on a branch high up, Brutus was a young Banyan, forced to go all-out in a fight for survival. A variety of strangling fig that overpowers its hosts, evidence of a Banyan's determination can only be marvelled at when hundreds of tourists a year visit the Acharya Jagadish Chandra Bose Botanical Garden in West Bengal to pay homage to India's national tree.

Spanning over 3 acres, the garden welcomes visitors into what appears to be an enchanting shady forest. Sunlight filters through dappled leaves, and clear sky glimpses through lofty boughs. Look closely, and you'll discover that the entire garden is one mighty Banyan tree, over 80ft tall and comprising 3,000 aerial roots, making it the largest tree canopy in the world. Surviving cyclones, famine and civil war, and despite having some of its main trunk removed in the 1950s due to rot, this fearless warrior's unrivalled resilience lies in a genius triple-defence strategy – buttress, stilted *and* aerial roots.

In an early courtship that seemed harmless enough at the beginning, baby Brutus sent ever-thickening roots downward that eventually rooted once they hit the ground, encircling his foster parent in a smothering embrace. Before the tree knew what was happening, it was forced to compete for sunlight and sustenance, growing increasingly wary of its uninvited lodger. Not content with choking the life out of their hosts, Banyans like Brutus produce trails of dangling, vine-like roots, lassoing neighbouring trees to gain an uncompromising foothold. Such guerrilla tactics, combined with a suffocating web of massive ground-level roots, means over time Banyan hosts are forced to capitulate. But in nature, death to one provides opportunities for others as animals, birds and reptiles set up home in the decomposing structures.

You'd imagine that Brutus's formidable manoeuvres were enough to guarantee his supremacy, but the malevolent thug isn't done yet. He wages war in more subtle ways, relying on tiny fig wasps to fertilise its fringed, sweetly scented ivory flowers. Now a young adult, Brutus is not giving those up without a fight either and the relentless warrior plays his trump card. Forced to crawl inside a tiny hole in his figgy fruit, losing their wings and antennae in the process, male and female wasps are immediately deprived of any physical means of escape. With little else to do but mate, females lay eggs in the flowers, scattering pollen that fertilises Brutus's tiny blooms. But, driven by a primitive need for survival, Brutus still demands his pound of flesh. Instead of gratitude, the fruit digests the dying wasps and their eggs hatch, giving birth to a new generation of mating wasps. Trapped and with no way out, shackled males bore holes in the fruit, allowing starving females to escape to collect pollen and abandon their male counterparts to eternal damnation.

Have you noticed human relationships often play out the same way? That charming guy or girl who seemed such fun and so generous in the beginning slowly worms their way into your affection, stealthily switching from buying you dinner to letting you pay. At first, you think it's cute. Hell, it's only a few pounds or dollars here and there. And besides, you're in love! What the heck! Requests to borrow money for small things become increasingly bolder until suddenly you're bankrolling their lifestyle at the expense of your own.

Botanical pirates vary widely in their opportunism; some are mildly scrounging, while others shamelessly suck the life out of others with no intention of lifting a finger to contribute to life's daily trials. We'll be meeting some of these intriguing freeloaders later, but it brings me neatly to one of the most fascinating aspects of roots – symbiotic relationships.

Symbiosis is a wonderful thing, and like our own friendships, organisms have developed a mutual understanding of another's needs and are arguably more reliable in meeting their obligations than many of us. Living much of their lives underground, fungi lack an indispensable green pigment, chlorophyll, so they are unable to photosynthesise through sunlight. Instead, they develop slinking bootlace extensions that travel for miles scavenging for water and food, as well as essential nutrients like nitrogen and phosphorus. These resources are generously gifted to their hosts, thereby expanding their inherent capacity for absorption, and in return, fungi receive a form of glucose in payment. A win-win for both parties. It's easy to compare the connections many plants have with fungi to human ones. Sometimes, both the plant and fungus get something worthwhile out of the relationship, and at other times, it can be distinctly one-sided.

Mycorrhiza is a hot topic at the moment; something you might well have spotted at your local garden centre. You can't buy a plant nowadays without being offered a pouch of mycorrhizal plant food. Basically, it's peddling plant empathy in a packet, a formula that promises to help roots establish better and maximise nutrient absorption. I'm not convinced, and neither was Which Gardening, which tested a number of products on the market. Given that around 85 per cent of plants are naturally mycorrhizal, soil already contains a wide range of beneficial fungus that allow plants to enjoy mutually beneficial relationships at root level. Testers found that sometimes these products worked, but more often than not, they offered no real advantages at all. Touted as the new cure-all, in my humble opinion, improved soil fertility is guaranteed by simply adding plenty of organic matter. A solution far cheaper, longer-lasting and effective than so-called magical mycorr pouches.

We can't continue our globetrotting without meeting the world's most expensive fungi – truffles. Globally recognised as the gold standard of gastronomy, top chefs and private gourmets are prepared to pay as much as €3,000–4,000 per kilo for coveted black Périgord truffles. Traditionally, truffle hunters used female pigs, which are drawn to the truffle's musky scent because it mimics a sex pheromone of male pigs. Dogs have replaced pigs, and today, the global truffle market is worth £270 million with demand expected to rise to £630 million by 2033.

Over 70 per cent of the world's truffle supply comes from family-run businesses in northern Italy, Spain and Mediterranean France, where oak and hazel woodlands with moist, well-drained alkaline soils are common. Rainfall is vital for good truffle harvests; however, increasing drought threatens future yields in traditional growing regions. It makes me think that the UK's changing climate, with warmer summers and increasingly wetter winters, could prove ideal for truffle growing. The late Prince Philip, the Duke of Edinburgh, planted 300 oak and hazel saplings inoculated with truffle spores at Sandringham, hoping to produce a debut British truffle. Twelve years later, his patience paid off with the estate's first black truffle harvest. Don't expect quick profits though; after young oak and hazel trees are inoculated with truffle slurry, it takes five or six years to harvest 'dirty diamonds'. But in a global market where demand outstrips supply, it could prove a canny investment for patient panhandlers. However, plant hunters of old had little truck with patience and set off adventuring across unchartered terrain to discover untold floral treasures that we now take for granted.

PLANT HUNTERS: Curiosities and Commerce

Intrepid plant hunters have done their fair share in spreading plants across the globe. We find one of the earliest examples in ancient Egyptian ships sailing from the legendary land of Punt. And in what is considered one of the earliest examples of plant hunting, this was no ordinary expedition.

Let's travel back in time and meet the young queen Hatshepsut, who harboured a secret ambition to rule. Married to and co-reigning with her 12-year-old half-brother, Thutmose II, her opportunity arrived upon his death. Said to have been aided by her lover and chief minister (who doesn't enjoy a bit of political intrigue?), Hatshepsut seized the throne, declaring herself Pharaoh in place of her stepson, the rightful heir, who was nearing maturity. Once in power, she revealed a keen head for commerce, with her sights set on an even greater prize: frankincense and myrrh. With a value equal to and often greater than gold, these oils were deeply prized in Egyptian culture. Rather than sending armies to war, Hatshepsut dispatched soldiers on a grand expedition to Punt to secure domestic supplies. Laden with gold, ivory and exotic livestock, the returning ships also carried over thirty valuable saplings of frankincense and spiny myrrh, their roots carefully packaged for the voyage home – a crop destined to bring untold wealth and prosperity under her reign.

It's hard to believe that spices and oils that are now relatively inexpensive and freely available were once worth more than their weight in gold. In the sixteenth century, English, Spanish and Dutch sailors plundered the West Indies for nutmeg, a commodity that was far more valuable than gold. Ransoms were paid in peppercorns, and Roman soldiers received their wages in salt – giving us the word 'salary'. Even today, saffron harvested from crocus stamens commands an extraordinary price, and in 2007, the record sale of a 1.3kg Italian white truffle reached £160,000 in a heated bidding war.

I daresay Italian-born Christopher Columbus could scarcely have imagined the sensation caused by a single pineapple, which he brought back from the Caribbean in 1496. Rare, exotic and nigh on impossible to cultivate in England's mercurial climate, pineapples had become status symbols by the mid-eighteenth century. Too costly to eat, they were instead used as opulent table displays at grand banquets, and those unable to afford them could rent a pineapple for special occasions. Importing such exotic fruit from the colonies without spoilage was a challenge, yet wealthy landowners spared no expense, hiring gardeners to stoke special pineapple pits heated with steaming horse manure to coax the fruit to grow.

Though the term 'plant hunter' had not yet been coined, early missionaries and medieval monks would have been well positioned to exchange botanical knowledge. By the eighteenth century, priests sent abroad to convert non-Christians discovered new floral curiosities, which they carefully sketched, painted, recorded and sent home.

The Victorians were obsessed with rare plants. Wealthy patrons funded costly expeditions to bring home exotic species. From Africa, Asia, the Americas and Australia came countless plants, trees and shrubs – so familiar today that it's hard to imagine they were once rarities. Modern gardens are filled with treasures carried back centuries ago in an era when industrial innovation and social upheaval fuelled a Victorian frenzy for horticultural novelties. While cottage gardens had long sustained rural peasants, ornamental gardens – traditionally the preserve of the landed gentry – were suddenly emulated by the nouveau riche eager to outshine the old money with botanical status symbols.

Illustrious figures such as Charles Darwin and Captain Cook's adviser, Joseph Banks, returned from Botany Bay with over 1,300 rare plants, helping to fuel this mania. Glasshouses were constructed to protect delicate exotic specimens, igniting fierce competition to showcase the newest floral wonders. Alliums, bamboo, camellias, acid-yellow spurge and vibrant rhododendrons are just a few of the many plants that successfully emigrated and now flourish the world over, thanks to the daring of plant hunters.

Seed banks, meanwhile, preserve centuries of collected biodiversity. London's Kew Gardens safeguards a dizzying 2.5 billion seeds from around the globe, stored in controlled conditions for use in science, medicine, experimentation and rewilding. But plant hunting isn't always a benign pursuit. In 2000, British horticulturalist Tom Hart Dyke and his friend, merchant banker Paul Winder, embarked on an orchid-hunting expedition in Colombia's rainforests – then vanished without explanation. After weeks of fruitless searching, their families feared they had been killed. The truth was stranger: captured by guerrillas, the pair were held hostage. Beaten, tortured and fed spiders, snakes and monkeys, they were forced to write ransom notes for $5 million each – letters that never reached their intended recipients. Against all odds, and after nearly ten months in captivity, they were released as abruptly as they had been seized. Years later, Hart Dyke returned to Colombia at the invitation of the Ambassador, though wisely limiting his search to nurseries rather than jungles.

As he flew across the Caribbean and Pacific coastlines, Dyke may well have reflected on wider truths: that plant hunting is only one piece of a vast ecological puzzle, and that the marine environment plays just as vital a role in sustaining life on Earth.

THE PLANET'S LUNGS:
Every Breath You Take

Our voyage of discovery will take us across the globe, but for now, let's immerse ourselves in the balmy Mediterranean. Warm sunshine caresses your face, your toes sink pleasurably into damp sand, seawater laps at your feet – you relax and draw in a deep breath. Equally seductive are the waters of the Sargasso Sea, where gentle waves wash pink, sandy beaches and scour hidden Bermudian coves, while iridescent eel and tuna dart through sparkling turquoise depths. In the shallows, crabs and hatchling turtles feed on sargassum – a nutrient-rich, free-floating seaweed for which the sea is named. Unlike most seaweeds that anchor to the ocean floor, free-spirited sargassum reproduces vegetatively, its leaves equipped with air-filled pods that keep it afloat. Enriching otherwise barren waters, this remarkable seaweed provides vital food and shelter for humpback whales and sharks, and marlin that return here annually to spawn. But what the Sargasso Sea and the Mediterranean share is nothing short of miraculous; both are home to the smallest known organism on Earth – a seemingly unremarkable green microbe with an unpronounceable name: *Prochlorococcus*.

Thriving in the sunlit zones of temperate and tropical oceans, this microscopic, plant-like organism uses photosynthesis to generate energy and plays a critical role in sustaining marine life. Here's the astonishing part: *Prochlorococcus* produces around 20 per cent of the world's oxygen – more than all the tropical rainforests combined. That deep, calming breath you just took seems all the more miraculous now, doesn't it?

As impressive is phytoplankton, microscopic algae that generate roughly 50 per cent of the Earth's oxygen while transferring vast amounts of carbon dioxide from the atmosphere into the ocean.

Like trees, phytoplankton capture sunlight through chlorophyll, absorbing carbon dioxide and releasing oxygen. Under the right conditions, their populations explode in blooms that are visible from space and cover hundreds of square miles for weeks, although each organism lives only a few days. Once they die or are eaten, their carbon either sinks to the seabed or is recycled near the surface. While some species trigger toxic red tides that can devastate marine life, phytoplankton remain the foundation of oceanic food webs, feeding small fish and invertebrates.

And then there are whales. Majestic leviathans whose very existence strengthens this massive carbon pump. Being warm-blooded mammals, whales nurture their young in the womb, breathing air through nostrils that through the ages have evolved into blow holes. Over millennia, their bodies have transformed: hair has been replaced by smooth, streamlined skin, limbs reshaped into fins and tails into powerful propellers. As they roam the oceans devouring krill and fish, they surface for air and to defecate, releasing nutrients that feed phytoplankton. In turn, phytoplankton fixes an astounding 30–50 billion metric tonnes of carbon every year.

It's only in recent years, as climate concerns mount, that I've started paying close attention to these details. Yet, for all the doom-laden headlines, the resilience of plants – whether on land or sea – remains a miracle. They provide the air we breathe, enrich our ecosystems and stand as our allies in the fight against climate change.

Of all marine ecosystems, coral reefs may be the most diverse, sheltering a quarter of all ocean species while supporting countless aquatic plants. Coral reefs lure tourists, divers and sports fishermen, boosting economies while providing livelihoods for coastal communities. And guarding these precious coastlines, a group of unsung heroes stand firm: mangroves. These hardy trees form natural storm barriers, stabilise shorelines, filter pollutants and improve water quality. When seas rage and storms roar, believe me, mangroves are the guys you'd want on your side. To meet them, we must journey to the tropics.

Compared to the dazzling rainbow of coral reefs, mangrove forests appear ghostly and tangled, their propped roots rising eerily from brackish waters. But what they lack in glamour they more than make up for in resilience. Enduring relentless heat, salt and shifting tides, mangroves have evolved efficient salt filters and specialised aerial roots. These adaptations underpin their distinctive anchorage, allowing them to thrive where most plants would quickly perish. And as if that wasn't enough, their strong root networks slow water flow, providing nurseries for fish, rays and sharks and trapping pollutants.

Alongside mangroves, humble seagrass – Earth's only flowering plant able to thrive in pure seawater – plays a front-line role in defending our coastlines. From craggy Cornish cliffs to breathtaking Scandinavian fjords, seagrass meadows are vanishing, prompting urgent restoration efforts by groups such as the World Wide Fund for Nature (WWF). And no wonder. Seagrass absorbs a staggering thirty-five times more carbon than rainforests. Just a single 10,000m^2 meadow can support over a million invertebrates and 80,000 fish and stores up to 18 per cent of all oceanic carbon, despite the fact it covers less than 0.1 per cent of the sea floor. Restoration brings swift reward. In Sweden, 80 per cent of invertebrates returned to new meadows within just three months, and closer to home, volunteers on Britain's south coast recently planted 70,000 seagrass seed bags to restore 3.5 hectares of seabed.

Back on dry land, plants continue their quiet miracles. Pollen-bearing bees and butterflies ensure the next generation of crops and flowers, young saplings grow into forests, seeds into food and nectar into honey. The American philosopher Ralph Waldo Emerson wrote, 'I declare this world is so beautiful that I can hardly believe it exists.' Which is all very well, but I guess we should talk about plastic.

PLASTIC POLLUTION:
Plastic, Not Fantastic

Plastic was a game-changer. The wunderkind of its day, revolutionising industries from manufacturing to packaging, medicine to technology, and first introduced in 1862 by English-born chemist Alexander Parkes. It must have been thrilling when, prompted by concerns that elephants would face extinction if their tusks continued to be harvested irresponsibly, he patented his new material, Parkesine, a cheap alternative to ivory and tortoiseshell, which was used widely for combs, billiard balls and piano keys. By 1902, Belgian chemist Leo Baekeland had developed the first fully synthetic plastic using heat-treated formaldehyde and phenol. This new material – later named Bakelite – ignited a consumer craze for stylish art deco designs, quickly becoming integral to iconic products such as ashtrays, cameras, radios and telephones. With the rise of Hollywood in the 1920s, celluloid became the standard for cinema film and movie stars sported sunglasses, sparking a social trend for fashionable sunshades.

No one can deny the sheer genius of plastic innovation. Cheap, versatile, durable and easily moulded, it could be shaped into endless forms. Remarkably, during the test launch of the atomic bomb in New Mexico, support troops were offered only plastic sunglasses and sunscreen as protection against radiation! The Second World War not only accelerated plastic adoption, but some argue that plastic was as crucial in securing the Allied victory in 1945 as military dominance itself. Shortages of natural resources pushed the United States to seek cheaper synthetic substitutes: nylons replaced silk stockings, and steel, wood, paper and packaging were quickly superseded by low-cost plastics.

Then along came a more enhanced plastic resin – PET – which opened the doors for an even wider range of everyday products, from packaging, containers, clothing, automotive design and electrical components, trouncing earlier plastics in its sheer versatility. Lightweight, waterproof and practically unbreakable, world demand surged for cheap products – a trend that has continued unabated. For all its innovation, unfortunately, plastic lingers.

Today, the plastics industry relies heavily on fossil fuels, making it a major contributor to global carbon dioxide emissions. It wasn't until the 1960s that concerns about its environmental harm gained traction, as scientists realised plastic doesn't disappear but instead breaks down into microplastics. American folk singer and activist Pete Seeger was surely something of a visionary in his warning, 'If it can't be reduced, reused, repaired, rebuilt, refurbished, resold, recycled, or composted, then it should be restricted, redesigned, or removed from production.'

Ironically, the very qualities that made plastic so appealing – durability, low cost and ease of production – are the reasons it has become such a profound environmental threat. Yet, in response, a new revolution has begun; one seeking sustainable alternatives inspired by nature itself. Which brings us back to that little powerhouse, *Prochlorococcus*, the giver of every one in ten breaths we take. Alongside algae and plants, these organisms form the lungs of our planet, generating much of the oxygen we depend on. Yet plastics and other pollutants are undermining this delicate balance. A recent study by Macquarie University in Sydney revealed that harmful chemicals impair *Prochlorococcus*, altering hundreds of its genes. But with so little research into how plastic pollution affects marine organisms, this may be only the tip of the iceberg, raising stark questions about the consequences for humanity if these ecosystems were to collapse.

Marine life already suffers gravely. Countless creatures are injured, entangled or killed by ingesting plastic debris. The infamous floating plastic island in the North Pacific – half of it made up of synthetic fishing nets – is a shuddering symbol of our throwaway culture. Shockingly, 80 per cent of ocean plastic comes from land-based sources, with the remaining 20 per cent from marine activity. Most plastics, being non-biodegradable, circulate endlessly in the oceans, creating a toxic swamp of bottles, bags, nappies, flip-flops and fishing gear. It's estimated that as much of 70 per cent of this detritus sinks to the ocean floor, and combined with oil and gas run-off, the picture resembles a slow-motion global disaster.

The fragility of our planet is becoming increasingly difficult to ignore. As Barack Obama cautioned in April 2023, 'We are the first generation to feel the impact of climate change and the last generation that can do something about it.' Governments, industries and individuals are awakening to the need for sustainable choices because any nation that destroys its natural resources is on a collision course with self-destruction. Still, there is hope – if we act.

Let's imagine for a moment that plastic could be good. What if we could create a material as versatile as plastic that is fully biodegradable? Humans created plastic – and that same ingenuity may yet create something better, paving the way for a future where progress doesn't come at the expense of the planet. We generate around 422 million tonnes of plastic annually with 5 trillion plastic bags used each year, half of which are discarded after a single use. Despite recycling efforts, only 9 per cent of plastic is recycled; the rest pollutes land, water and food chains.

Finding an alternative is no small feat, but innovation is already under way, and thanks to some exciting breakthroughs, we've got good reason to feel hopeful. Bioplastics, often made from plant cellulose, are continually emerging. A London start-up is developing seaweed-based films; Harvard researchers have created biodegradable plastic from shrimp shells that decomposes within weeks not centuries; and pineapple fibres are being transformed into stylish trainers. Nature itself offers solutions: mealworms can survive on polystyrene; darkling beetle larvae can digest foam; Japanese scientists have discovered bacteria that consumes plastic bottles; and researchers in Texas have engineered an enzyme capable of breaking down PET.

And while humans scratch their heads over environmental issues, plants are constantly innovating with their own unique responses to life's challenges.

PART TWO

SENSES AND DEFENCES

Necessity hath no law.

– Oliver Cromwell

Plants may look passive, but they have developed remarkable ways to protect themselves. Many grow thorns or spines to make grazing difficult, while others fill their leaves and roots with chemical toxins. Some plants cleverly disguise themselves to blend into their surroundings and a few even call for reinforcements by harnessing hungry insects to chase off their attackers. Though rooted in the soil, they are rarely helpless.

ANXIETY: Burnout and Stress

Accepted wisdom often advises us to go with the flow, to weather every storm life throws at us. Tangled love lives, financial stress and emotional upset, vampire bosses and impossible workloads, every new challenge weighs us down. We bend, we bear it, until one day – we break. And similarly, so do the plants around us. Aesop's fable of 'The Oak and the Reeds' recounts the tale of a mighty oak growing beside a stream, stout branches reaching for the sky, so that when the wind blew, it mocked the shivering reeds below. We all know the ending – when a fierce hurricane blew in from the north, uprooting and toppling the arrogant oak, it was the flexible reeds that survived.

Exposed to blistering winds, sleet, snow, ice, drought and scorch, it's no surprise that plants suffer anxiety, stress and shock. If wounded, they can often (although not always) heal, returning to life once the current threat recedes or the growing season is under way once more. Every year, branches are split by lightning or strong winds, trees are uprooted, leaves crisp under a blazing sun or suffer soggily from excessive wet. Plants are dug up by animals and other scavengers or tossed on the compost heap because a gardener couldn't tell a weed from an ornamental. They stand exhausted, silently begging for rain or warmth that may or may not come. Colours fade to brittle shades of brown and yellow, dropping their leaves, unable to sustain the effort that keeps them working fluidly.

Plant stress can be long lasting, recurring or temporary. Unsurprisingly, long-term stress is harder to combat than temporary pressures but those unfortunates that find themselves forever mired in waterlogged soils or stranded in compacted soil exist in the never-ending misery of depleting nutrients and oxygen, which, without respite, presents life-threatening conditions. Naturally, moisture lovers such as loosestrife, swamp cypress, willow and hart's tongue fern are well used to having their roots puddled in water for extended stretches and can shrug off seasonal flooding or temporary waterlogging. But it proves a catastrophe for plants growing in the wild or agriculture, when prolonged flooding can wipe out an entire potato crop within twenty-four hours. Climate change has increased rainfall and flooding in vulnerable areas and for many plants, waterlogging is fast becoming one of the greatest threats they face. At other times, stress is temporary, caused by immediate impact from brief drought, hosepipe bans, flash flooding, temporary freezes, construction damage or wayward parenting.

To wind that back to a more relatable setting outdoors. If you're the type of gardener who has a tendency to overwater, it can prove just as harmful as water deprivation. Dark spots, pimples and corky patches on your garden plants are a way of telling you they are drowning, not waving. House plants are vulnerable to drunken stupor, with unwitting owners killing them with kindness. It's generally accepted that houseplants pretty much thrive on neglect and if you overwater your charges on a regular basis, their leaves just can't cope with the amount of water the roots send up. Subsequently, cells rupture, roots begin to rot, while above ground, flowers drop and leaves begin to look decidedly off-colour before the whole thing withers and dies.

Indoor plants are frequently forced to cope with widely fluctuating temperatures that causes temporary stress, which, if left untreated, results in probable death. Tender plants, such as pelargoniums or bougainvillea grown in conservatories or those perched on windowsills, can drop their flowers and curl inwards, whispering of too many scorching afternoons pressed against searing glass. By dusk, they flop, sighing in relief as the sinking sun offers welcome cool. But then night settles in. Cold breath huffs at the windows, shivering their stems. By morning, their yellowed leaves drop and surrender to the relentless seesawing of heat and chill. Exposure to widely fluctuating temperatures is guaranteed to ramp up plant stress to unmanageable levels. So, be a good foster parent – move them out of bright light and nasty draughts to a more genial spot with diffused light levels and a steady, equable temperature.

Of course, the opposite can be true. During a parched summer you'll see foliage brown, flowers drop and stems swoon in the heat. All symptoms of an equally formidable foe – water deprivation. Gardeners often fret over their lush emerald lawns getting scorched during extended drought, thinking they will never recover – a temporary trauma that is easily and quickly remedied in a domestic setting with generous watering, ideally in the evening or early morning. Thankfully, grass is battle-hardy and while prolonged hot, dry weather might kill off top growth, nine times out of ten, roots motor on regardless, standing by to send up fresh growth once seasonal rain returns. All of this is proof that plants aren't as defenceless as we might imagine and, when cornered, they fight back.

Never did I envisage that bananas might influence global farming or agricultural practices outside of their local plantations. Well, you learn something new every day. Would you believe that many plants are able to sense approaching oxygen deprivation and defend themselves against it? In the advent of flooding, some plants produce ethylene, a gas with which most of us are familiar as it causes bananas to ripen. Sensing environmental change, bananas begin to accumulate reserves of ethylene, which generates another clever protein (similar to haemoglobin) that reduces levels of yet another gas, nitric oxide. Astoundingly, this signalling warns a plant to prepare for asphyxiation. Although different crops vary in their ability to detect impending disaster, it's a pretty staggering discovery, and understanding how plants react to flood risk provides a significant step forward in helping farmers breed crops with enhanced resilience.

For every action in life there is a consequence – the old cause-and-effect routine – and nowhere is it better illustrated than in the plant world.

CHEMISTRY: Chain Reactions and Chemical Warfare

Throughout history, humans have sought to unravel the mysteries of the world, transforming the unknowable into the knowable and devising tools to ease industrial, agricultural and everyday tasks. This relentless pursuit of knowledge and efficiency reached a turning point in the mid-eighteenth century with the Industrial Revolution in Britain, a transformation that swiftly spread across the globe and reshaped the way we lived. Horse-drawn haulage, farming and handcrafted industries were haltingly replaced by emerging technology. Gas street lighting came of age, along with the invention of the steam engine, the cotton gin and power loom. From whimsical, self-wiping napkins to hilarious automatic page-turners and beer dispensers, these quirky contraptions, celebrated for their humour as much for their ingenuity, captured the essence of playfulness wrapped in utility. As children, which of us weren't intrigued at perpetual motion in mechanical games like Mousetrap? Watching the silver ball bearing roll into action, setting off the crank that turned the lever which sprang the trap was pure magic. Though not a chain reaction in the strictest sense, it encapsulates the wonder of progressive events – a series of interconnected actions spiralling toward a satisfying finale.

Central to this inventive spirit is an enduring fascination with chain reactions – intricate sequences where one action triggers another, often with delightful or unexpected results. As we've discovered, the demise of the dinosaurs occurred through a sequence of disastrous climatic factors, and today, seismic events are still occurring. Off the coast of Sendai, Japan, in 2011, the rupture of the sea floor unleashed a tragic domino effect that left an indelible stain in modern history. The resulting earthquake (the largest ever recorded) triggered a cataclysmic tsunami, which in turn led to the Fukushima nuclear disaster.

Arguably, the most profound chain reaction in human history was unearthed much earlier, in 1938, by the German chemist Otto Hahn. Working alongside a team of brilliant physicists and mathematicians and awarded the Nobel Prize for Chemistry in 1944, Hahn discovered nuclear fission, a process as terrifying as it was ground-breaking. A discovery that set the stage for the Manhattan Project and, ultimately, the world's first atomic bombs being dropped on Hiroshima and Nagasaki in the final days of the Second World War.

While nuclear fission symbolises man's capacity for destruction, nature has long relied on a similar kind of sequence with far more favourable consequences: photosynthesis. Through this elegant process, plants convert sunlight into life-sustaining energy, and what happens to that energy is remarkably similar to a chain reaction.

Thanks to the light-absorbing pigments in their leaves, plants are masterful collectors of solar power. And much like solar panels, which rely on two silicon layers, one electron-rich and positively charged, the other negatively charged, plants create an electric field that enables the flow of electricity powered by sunlight. They convert this solar energy into usable forms of glucose and oxygen with astonishing efficiency.

It's easy to understand this function in the modern world. However, a much-overlooked historical footnote is accoladed to English chemist and religious maverick Joseph Priestly, who made one of the most important discoveries ever – oxygen. The early Greeks had already distilled the four unique elements of creation: air, earth, fire and water. However, in 1774, Priestly identified oxygen as a definitive gas for the first time by trapping a mouse in a jar with a lit flame. Observing that both the flame and mouse died, he diced with the odds by exposing the jar to sunlight and added a mint plant.

The flame burned longer and the rodent lived longer. Reasoning that the interaction of the plant and sunlight were somehow connected, and that together they were replenishing the air inside the jar, Priestly concluded that the mint must be releasing some sort of gas – oxygen – into the air, and the process of photosynthesis as we know it was officially identified for the first time.

In their battle for survival, plants have developed their own version of weaponry – allelopathy, or chemical warfare to you and me. A phenomenon where they produce toxic chemicals to nobble their rivals showcases nature at its most resourceful and ruthless. Plants release poisonous compounds that can benefit, suppress or kill their competitors and they deliver them with staggering precision. These fatal toxins are released as chemical gases and are absorbed by neighbouring rivals through various methods: via roots leaching poisons into the soil or decaying leaves that transfer toxins to nearby plants.

Far from Arcadian in its community relations, the tree of heaven employs a forensic finality in preventing its rivals from establishing by leaking poison through its bark, leaves and roots. Similarly, black walnut, with its plump, delicious nuts rich in antioxidants, packs its buds, leaves, husks and roots with juglone, a toxin that causes severe respiratory failure and suffocation in its competitors.

Other plants wage equally deadly warfare. Peas produce pisatin, a growth inhibitor that stunts surrounding crops such as cress and lettuce. Meanwhile, alfalfa, barley, rice and sunflowers release allelochemicals to suppress weed growth, giving them free rein to available sunshine, moisture and nutrients.

In the quest for more sustainable agriculture, nature's firebrand of chemical warfare both brutal and brilliant has opened doors to more sustainable future farming practices on a wider scale. Understanding these interactions has prompted farmers to employ plant-derived compounds to suppress weeds and influence crop and pest interactions. Similarly, the miracle tree, which has earned a reputation for prolific seed dispersal and rapid growth (it easily grows 10ft a year), is famous for its selective toxicity. An invidious poison spares its own species and seedlings with a unique nepotism favouring its success above all others – reinforcing the old adage that blood is thicker than water. Widely used to create shelter belts for plantation crops like cacao, coffee and tea, its resilience and adaptability made the miracle tree an excellent candidate for rehabilitating eroded and degraded lands.

Even innocent-looking garden favourites like French marigolds exemplify this clannish utility by releasing a compound that prevents threatening larvae from hatching which might otherwise eat its roots, making it a valuable companion in kitchen gardens or allotments. Later, we'll be exploring companion planting and how gardeners can leverage these ingenious mechanisms to optimise their growing spaces. One of the ways you can boost plant health is simply by talking to them. Sounds crazy? I know loads of gardeners who talk to their plants, I'm one of them. But are they listening? And more importantly, can they speak to us?

COMMUNICATION: Plant Whispering

British readers of a certain age saw nothing unusual when Little Weed called to Bill and Ben in the BBC children's classic *The Flower Pot Men*. Sadly, we leave such childish imaginings behind, but if I asked you to step into your garden and bend your ear to a gaudy blossom or shrub, would you hear them talking? Naturally, weeping willows don't weep real tears, any more than wailing willows wail. Flowers don't cry out in pain when they are cut – or do they?

Without brains, emotions or neurons, let alone ears, eyes, mouths or noses, it sounds decidedly loopy to entertain the notion that plants can feel, hear, walk or talk. Yet, current science suggests it isn't as laughable as it sounds. Botanists, researchers and scientists have begun to probe plant intelligence believing plants can learn, react, remember and sense in ways that are familiar to us. Okay, plants are intelligent in ways we don't fully understand, but it does nothing to diminish a growing consensus that they react to chemical, sensory and acoustic stimuli.

In the mid-1800s, the notable British naturalist and keen botanist Charles Darwin had a head start on modern boffins. His seminal work, *The Power of Movement in Plants*, led him to explore plant behaviour. And since orchids were his particular penchant, he observed that orchid pollination was stimulated by the touch of an insect's proboscis and concluded that plants were sensitive creatures in possession of behaviours that permitted them to respond to the environment through sunlight, touch and gravity. Irrefutably, plants possess a subtle awareness, a sensitivity to the world's hidden cues. Roots snake under the soil, reaching towards traces of moisture. When faced with toxins, they veer aside, instinctively recoiling from danger as if drawn by invisible currents. Flowers pivot in slow, silent arcs, tracking the path of the sun across the sky, and even the slightest brush of a fingertip causes a shy retreat in some species. If you're sceptical of plant intelligence, studies conclusively prove they communicate.

A study at the University of Tel Aviv proved that plants *do* talk, and it seems to occur when they are thirsty. In a quiet laboratory, a soft crackle pulses from a microphone – faint, desperate pops. These were vocalisations of thirsty, drought-stressed tobacco and tomato plants, the sounds attributed to xylem tubes moving water and nutrients around the plant. Amazingly, those clearly audible thirsty pops were pinpointed with 70 per cent accuracy. Studies show that many plants, including cacti, corn, grape and wheat, emit stress cries too. Far from suffering in silence, drought-stressed and wounded plants with cut stems emitted up to thirty-five airborne sounds per hour versus quieter, well-hydrated, healthy plants making about one sound per hour.

Equally intriguing, a further experiment prompted researchers to introduce two female moths to two healthy tomato plants. One plant was kept in silence, while the other was rigged with a speaker that played sounds from a dehydrated plant. You don't need three guesses to work out which plant the moths preferred – the healthy, silent one! Perhaps bats or mice more finely tuned to ultrasonic sounds are perfectly capable of making sense of plant talk even if we struggle to understand it.

Lacking specialist sensory organs, it seems impossible that plants communicate, let alone generate unique sounds in their environment. Birds, frogs, fruit flies and snakes can sense sound without eardrums – so, why not plants? A growing body of evidence tells us they can. Plants can distinguish between chewing noises and sounds made by insects or wind. Further evidence from enterprising biologists from the University of Western Australia discovered that plants use sound waves to detect water at a distance and can home in on moisture when it's closer to them. Equally extraordinary, the pale, lemony flowers of beach evening primrose release sweeter than normal nectar on hearing the sound of buzzing bees and rock cress floods its leaves with defensive chemicals the moment it senses a caterpillar chewing on its foliage. Okay, they don't actually *hear* the bee or caterpillar, but it's thought the vibrations of a chomping herbivore are enough to signal a chemical response and, astonishingly, even an audio recording of munching caterpillars triggered the same defensive reaction.

As we've seen, prairie plants can sense chemical changes wrought by smoke, so it comes as no surprise to learn plants are equally capable of warning their neighbours when danger is imminent. It's a kind of neighbourhood watch, where they emit invisible compounds that drift through the air, sending an alert to nearby plants to gear up their defences. Unfortunately, plants aren't the only ones to hear the distress signal. Nearby, a huddle of beetles pause their steady crawl, sensing something is afoot. One by one, antennae twitch, and they trace the alarm, zeroing in on a wounded branch or plant where the action is unfolding.

Acoustics cause vibrations in plant tissue every bit as faithfully as sound resonates in the human ear. Plants experience good and bad vibrations from all around, both nearby and far away. Vibrations travel through the air, soil or via immediate stimulus when something such as water, stone, air, animals or a person make physical contact with a plant. Conversely, local low-frequency vibrations might come from moving insects – grasshoppers clicking in wild grass – or from natural forces like wind and rain. However, plants are also sensitive to high-frequency vibrations such as feeding insects or distant vibrations, including birdsong, buzzing insects and flowing water, and are highly adapted to distinguish the good from the bad.

In laboratory trials, two different soundtracks were played to two different sets of plants and elicited different responses. Plants subjected to an audio of feeding caterpillars ramped up the production of mustard oil, a natural insect repellent. The second batch were treated to a silent recording and didn't react at all. A second sound test upped the ante. Both batches were played the mating songs of leafhoppers and wind, and both sets failed to respond. So, it appears they know which vibrations are threatening and which are harmless and can be ignored. The wonders of green folk!

Even if humans can't appreciate it, insects have picked up on this auditory sensitivity. Leaf and treehoppers are smart enough to work out that plants can help them attract mates by sending vibrational mating calls along plant stems by snapping their abdomens like elastic bands. Masters of industry and communication, ants have developed a mutualistic relationship with a particular acacia in the African savanna, the whistling thorn – home to at least four species of ants who have been known to fearlessly attack a herd of elephants to defend it. Elephant trunks are especially sensitive and, alerted by the vibrations of grazing elephants, ants swarm to the tree's defence and attack at the first sign of trouble.

Fascinatingly, elephants are excellent long-range transmitters, stamping their heavy feet and creating a thunderous wave that can be received many miles away by different herds, communicating through a diverse array of complex calls and signals, each serving a specific purpose – alerts to family members in danger; co-ordinating group movements; resolving conflicts; or the amorous sequencing of females to attract mates. Typically, elephants produce over seventy distinct vocalisations and employ more than 160 gestural, tactile and visual cues in their daily interactions to express their needs and strengthen family bonds.

Male hummingbirds vibrate their tail feathers in an audible melody in an energetic mating dance guaranteed to attract females. Bats use reverberating ultrasonic squeaks that bounce off trees and other objects to navigate during their nocturnal hunting. Baby gorillas beat their chest at play, while posturing adults thump impressive chests, sending vibrational messages over the miles. Inventively, a snake's inner ear is attached to its jawbone, which allows it to sense the movement of small rodents through vibrations running along the ground. And our slithering, scaly friends have more in common with whales than you might imagine.

Messaging by unfathomable means, including pressure waves and particle motion, whales also have ear bones fused to their skulls, suggesting these giant underwater mammals might conduct sound through their bones. Exactly how is something of a mystery since the complexity of their communication and behaviour remains as uncharted as the deep waters in which they swim. Given that vibrations travel though water much faster than air, it also explains the entrancing symmetry of shoaling fish, each mirroring those mercurial sharp twists and turns. Hearing organs hidden in their heads let them pick up vibrational sound waves, whereas cod make low, drumming sounds emanating from their swim bladders to communicate with their shoal-mates, and lumbering hippos can sense low-frequency vibrations underwater, even when their heads are above the waterline.

Back in the 1980s, HRH Prince Charles, then a king-in-waiting, admitted to talking to his plants. While it was sniggered at by the media at the time, His Majesty had the last laugh. Both NASA and the Smithsonian Institute have demonstrated that plants grow quicker in response to mild sound vibrations than stronger ones, so talking to your plants in a gentle tone encourages better growth and, would you believe, improves their immunity? Those of you who chat away merrily to your plants are officially vindicated from being called crackpots – hurrah! But if you *insist* on exposing your green charges to your playlists, plants are more likely to respond favourably to soothing ballads and love songs than, say, Linkin Park or Metallica.

Astronauts, of course, have been getting musical wake-up calls since the launch of Gemini 6 back in 1965. American astronaut Peggy Whitson, one of the best-travelled astronauts of all time, having accumulated over 650 days in space, must have listened to all kinds of music up there to alleviate the monotony, and famously quipped, 'After floating for nine and a half months in space, gravity is not your friend.'

Plants beg to differ.

GRAVITATION: Defying Gravity

On a beautiful English afternoon in the late summer of 1666, Issac Newton saw an apple fall from a tree in his mother's garden at Woolsthorpe Manor, Lincolnshire, and wondered why it didn't it fall sideways or float upwards. That falling apple gave rise to his gravitational theory that all physical objects, be they peaches, people or planets, exert a force on other physical objects, either pushing or pulling in a certain direction. He reasoned that the bigger the object, the stronger the force, and concluded that there must be different kinds of forces.

That famous apple birthed the notion of gravitation for the first time, launching Newton's reputation. We don't hear much more about that fabled tree, but the story goes that although Newton failed to mention the species, a visitor to the family home noted that there had only ever been one apple tree growing in Mrs Newton's garden. Damaged in a storm in 1816, the tree, minus many of its limbs, stood firm for another 350 years, and although the original is no longer with us, you can visit a commemorative one – a grafted descendant – at Trinity College, Cambridge, which reminds us of one of the greatest scientific discoveries ever made.

Newton was a marvel in other areas of plant science, too. Back in the eighteenth century, even the poorest and uneducated had worked out water flows downhill, a phenomenon that fascinated Newton, who pondered on how over a lifetime, trees could move millions of gallons of water a day from their roots to the highest boughs. Defying gravity, a tree of around 100ft can lift several tonnes of water a day. And as it turns out, water transportation has less to do with the tree itself than with the properties of water. Newton, with precious little research to fall back on, came uncannily near to the truth in unravelling this behaviour. Not with a 100 per cent accuracy but pretty damn close.

As we know, photosynthesis relies on water. He surmised that plant pores were full of water, which, when hit by sunlight, pushed water through the plant. And it's the hydrogen in water that goes on to form food (sugars) while the oxygen is exhaled through leaf pores. Evaporation naturally occurs through the leaves, and it is this gravitational pull that draws water from ground level up. Water molecules are sociable, able to stick to each other and other surfaces – while *adhesion* lets a raindrop stick to your window, it is *cohesion* that holds water together. So, with roots exerting their own pressure and forcing water through multi-channelled xylem highways, trees pull off this astounding triumph on a daily basis. If Newton hadn't tried to understand the science, I guess we'd still be thinking it was some kind of sorcery.

Plants sense gravity through sensitive cells in their root tips and will grow determinedly, no matter what. You only have to turn a plant pot on its side to prove the theory: it's known as *geotropism*. Cells sense a change in gravity and switch direction, so something is going on because the upper parts of the plants still grow in the opposite direction. Does that mean plants have two brains, one governing underground activity and the other directing above ground operations?

The elegant golden-limbed weeping willow that punctuates pastoral landscapes the world over and the graceful weeping silver birch both seem to defy gravity, with both trees growing in two directions at once. Their branches dip towards the ground and their trunks remain stiffly upright. This is a curious behaviour caused by a mutation in the WEEP gene that is still not fully understood. But it seems there's a greater expression of an extremely versatile hormone, auxin, at work that determines how branches display. Typically, auxins stimulate rapid cell division, controlling the direction of growth – a higher concentration of auxins at the base of a plant directs it to grow vertically, while greater concentrations of auxin on the upper side of shoot tips control the downward growth of weeping branches while keeping trees towering skyward.

You wouldn't think that air plants, with their distinctive wiry frizz, are related to Columbus's pineapple, and they seem to fly in the face of gravity by floating in mid-air. Unlike parasitic plants that live at the expense of others, epiphytes, as they are known, survive without soil or roots, tangling contentedly with other species without robbing them of life's essentials. Superbly adapted to life in the humid forest canopy, air plants simply latch onto a tree branch, absorbing vital nutrients and moisture from the air and rain through fine hairs on their leaves.

Another tiny pioneer, the humble desert moss is equally innovative in its fight for life, thumbing its nose at Newton's Law on a daily basis. Bypassing more traditional methods of collecting water at its roots, instead, it absorbs moisture from mist, fog or humid air from all directions, even able to perform this handy trick upside down.

Fine water-absorbing hairs at its leaf tips bind together to form droplets and send water sailing along micro-grooves to the leaves. With an astounding resilience to survive extreme habitats, desert moss has earned its reputation as a pioneer plant with the potential to enable the growth of other plants and perhaps transform uninhabitable barren landscapes, including space.

Wherever you look, you'll find rock cress of some description growing somewhere. In cracks in walls and pavements, on wastelands and rooftops, railway tracks and motorway verges, it is quietly conducting an agenda no less urgent than ours. But in 2018, it was about to get the lesson of its life. Putting rock cress at the forefront of its mission, NASA launched the first-ever designer batch of seed to be grown in a specialised laboratory on the International Space Station. Harbouring ambitions to allow astronauts to spend longer periods on Mars in the future, it is hoped that studying rock cress and how it reacts might provide insights into growing food for future astral journeymen. A never-ending galaxy lit with millions of stars makes us think about light, one of the most critical components in any plant's life.

LIGHT:
Responses to Light

Being sightless, plants don't see things exactly like us. Yet they are sensitive beings, acutely responsive to different types of infrared and ultraviolet light. You can test this for yourself with houseplants or cut flowers that lean towards the light, forcing you to keep rotating the vase so they don't grow lopsidedly.

As we know, plants use energy from sunlight combined with water and carbon dioxide to make sugars to power their physical functions.

And being a form of energy, light comprises a range of wavelengths that allow plants to sense different light levels through eyes of a kind. Photoreceptors scattered across a plant allow it to sense various wavelengths that govern its vital functions. When these receptors detect light, they send a signal triggering the production of auxins – that useful hormone again, which, among other things, effectively tells a plant which way to grow. To better understand how plants respond to light, we're travelling back to ancient Greece for a tragic tale of unrequited love, to a time when humans were forced to seek explanations for the universe through storytelling.

> Oceanus and his wife, Tethys, were high up in Greek society, and parents to a lovesick young water nymph, Clytie. Each morning, she would rise early to catch a glimpse of her heart's desire, the radiant Sun God Helios, as he raced his golden chariot across the sky, bathing the land in light. Sadly, the path of true love seldom runs smooth and heartbroken Clytie discovered that the dashing Helios was in love with another – Leucothoe, a mortal princess. Consumed by jealousy and determined to sabotage their romance, she revealed the affair to her rival's father, who was so enraged that his daughter should bring shame and dishonour on the family that he ordered she be buried alive. (An example of empathetic parenting!) On learning of her treachery, Helios shunned Clytie. Each day, she gazed longingly skyward, hoping he would forgive her and return her love. But Helios was in no mood for absolution. And our lovelorn nymph withered away from heartbreak and despair until finally transformed into a turnsole, a flower that follows the sun across the horizon. And so, the myth that sought to explain a sunflower's ability to follow the sun was born.

Driving through south-west France amid fields washed gold with sunflowers, you could be forgiven for thinking they are mimicking Clytie's pursuit of her hero. A poignant myth defrocked by science, but it's curious how close to the truth early Greeks were in their observations. Sunflowers are genetically programmed for solar tracking. However, contrary to popular belief, it's only budding sunflowers that turn towards the rising sun, following it until it sets in the west. Once night falls, sunflowers swing back to face east again, and once their flower buds burst open to reveal those gorgeous sunny faces, they stop tracking Helios, as it were, altogether and permanently face east. But scientists were puzzled at how sunflowers actually *saw* the sun to track it in the first place.

Almost all plants have the ability to grow towards light by means of circadian rhythms or, in our case, body clocks. A prayer plant in my kitchen of which I'm immensely fond supposedly folds its leaves for evening vespers, opening them again the following morning. Sometimes, for no reason at all, it seems to shiver, even when there is no air-con or discernible draught from an open window disturbing it. Popularly nicknamed 'the living plant', this endearing behaviour seems to be prompted by human-like circadian rhythms making its leaves move without apparent stimulus. In reality, it is simply folding its leaves as darkness falls, wisely conserving its energy for daylight hours when its mysteriously trembling foliage is tracking sunlight just like sunflowers (and I've also noticed it quivers with thanks when watered!)

Humans respond to the same internal chronometers, with darkness signalling it's time for sleep and awakening each day with the morning light. Circadian or biological rhythms also allow plants to anticipate and prepare for daily or seasonal environmental challenges and, remarkably, allow sunflowers to anticipate daybreak. An inner clock prompts them to synchronise the opening of their petals with the schedules of early rising pollinators – how typical that they should put fertilisation high on the agenda!

Equally, over-exposure to sunlight causes plants the same aggravation as it does humans – scorched leaves and sunburn – but instead of smothering themselves in Factor 50, sensitive-skinned plants have found a more attractive solution: silvered foliage. Originating in regions with sizzling summer temperatures, dusty miller, lavender, mugwort and olive possess appealing grey-green-hued complexions, a natural cuticle that acts as a natural sunscreen protecting them from scorch and also helps reduce water loss. As well as being a power source, sunlight or the lack of it persuades some organisms to adapt or position themselves more favourably. While land plants gain energy from sunlight, organisms living in the darkest depths of the ocean are often light-deprived. So, how do they survive?

Marine life is largely crammed into the top 200m of our oceans, where light is more abundant, and almost all marine plants and tiny organisms dwell in this 'photic zone'. Dive deeper, and they hit the 'aphotic zone', deep, dark waters where sunlight is minimal, with a weak blue-green light that prevents marine plants from photosynthesising. If organisms happen to live in the aptly named 'abyssal zone' – a darkness so complete that no sunlight penetrates – they are forced to feed on each other or any animal or plant waste filtering down from above. Yet, even without sunlight to make sugar, another miracle is taking place. Tiny microbes use chemical energy from hot water vents on the seabed to thrive, allowing deep-sea creatures such as fish, jellyfish, molluscs and crustaceans to create their own light through light-producing cells – bioluminescence – which allows even fathomless seascapes to glow. Think about sea turtles, who share an ancient heritage going back 2 million years with our earlier Stegosaurus. Despite having large, appealing eyes designed to see more clearly in bright conditions, they have surprisingly poor eyesight in dim water, but when waters get murky, this ingenious phenomenon allows them to pinpoint light-emitting prey with ease.

Back on dry land, all kinds of organisms glow in the dark. Millipedes, fungi, fireflies and endearing glow worms (that aren't worms at all, but beetles) harness their inner light for all kinds of pageantry. Femme fatale fireflies mimic the courtship displays of other species, luring unsuspecting males in with light fantastic, then devouring them. Decidedly more flirtatious female glow worms light up their beacons to attract mates, while their larvae flash to keep predators away.

And nobody really understands why the gills of Jack o'lantern mushrooms glow eerie green in the darkness, but should you want a novel fright night for the kids at Halloween, take them down to the woods at midnight for ghostly, glowing mushrooms and things that go bump in the night. Trust me, wind moaning through the trees, twigs snapping underfoot and the haunting hooting of night owls scares the bejesus out of them far more effectively than Casper costumes and witches' brooms.

If you have a nervous disposition, Halloween traditions might not be your thing. Fortunately, there are plenty of plants out there with gentler temperaments – the kind that amble rather than startle.

MOBILITY:
Plants in Motion

Treebeard of Fangorn from Tolkien's *Lord of the Rings* imaginatively defies reality. Or does he?

Life is often stranger than fiction, with nature providing us with, if not all-singing and dancing Ents, then certainly walking ones. Anchored by their roots, I don't need to tell you that real trees can't walk. Sure, they stretch upwards in search of sunlight or spread their branches sideways to distribute their weight, a prevention against being uprooted that allows them to bend and sway in strong winds, but that hardly qualifies as walking. Unless of course, you happen to be a walking palm.

Native to the rainforests of Central and South America, this unique palm would feel quite at home in Fangorn Forest because it's the only tree known to perambulate – an ability that is hotly contested by the scientific community, with some contending that its trunk remains fixed and the tree never moves at all, while others argue it can shift by as much as 20m over a couple of years. Growing up to 80ft tall, walking palms tower over their neighbours on gangling teepee roots that are perfectly propped to capture every inch of available sunlight. However, if the neighbourhood gets too crowded or soil is eroded at its roots, becoming nutrient-poor and threatening starvation rations, these short-distance strollers start their slow shuffle to greener pastures. Growing ever taller on stilts that shift slowly along the forest floor at a rate of about 2–3cm a day, as new roots pitch forward into richer soil, the older ones slowly rise up, seeming to facilitate the palm's 'walking' mode. Depending on what side of the argument you favour, the walking palm surely proves an exception to the rule because generally terrestrial plants simply can't move from place to place, right?

Think again. Geophytes – plants that have one seed leaf, such as bulbs – are blessed with underground storage chambers and can move to more promising spots, where they stand a better chance of survival, using specialised roots that shrink or swell according to environmental conditions. You can imagine there might be any number of good reasons why they might choose to up sticks and move on. Rain washes away surrounding soil, making life intolerable, rooting animals might dig them up, or perhaps sometimes they just need to seek kinder conditions away from severe heat or cold. Unlike their established elders, young bulbs can drag themselves through the soil, encountering less resistance by twisting and turning, often taking weeks or months to arrive at a spot more favourable to successful growth.

Much as I'm loath to admit it, under my wayward stewardship, my tulips are often forced to resort to similar measures. It's recommended that we plant tulip bulbs in November as they appreciate a cold snap to focus their energy, but with all the approaching Christmas shenanigans, I never seem to find the time to plant them. Through no fault of their own, they get shoved in willy-nilly on Christmas Eve, some carelessly upside down. (Admittedly, I've usually downed a glass of mulled wine or two by then.) Thankfully, they've never let me down yet, and my vertically challenged bulbs have the good sense to right themselves in time for spring blooming.

From walkers to runners. Plants have other methods of extending their territory. Strawberries produce long, pinkish, leafless stems, or runners, with infant plants at their tips that travel along the ground, sensing the ground under them and taking root. The gardeners among you know to wait a few weeks until they grow a few more leaves before cutting the umbilical cord between mum and child so you can safely dig up rootlings and pop them into pots of their own. (For the uninitiated, leaves, flowers and fruit will be identical to mum in every way.)

Mint, of course, is an infamous sprinter, happily swamping its neighbours with a robust root system that can grow over 1ft in a month. And even a common spider plant in the bathroom throws out gangly stems that if they could make contact at soil level would give rise to a whole new bunch of mini-mes. Talking of runners – videos of cats springing into the air at the sight of a cucumber proliferate online. Apparently, the mere sight of a cucumber is guaranteed to make your cat prickle in terror. Scotching the rumour that felines are cucumber-phobic, it's believed that it's the sudden appearance of an alien object that startles them, not because they mistake them for snakes, as commonly thought. But cucumbers can climb. They are superb cragsmen, latching coiling tendrils onto any handy vertical support, lifting them upwards. But the cellular biology explaining *exactly* how they do it remains something of a mystery.

Remember Brutus? His altogether better-mannered relative, the glossy Swiss cheese plant growing in your hallway, starts out in life much like its despotic cousin. Deprived of light and nutrition and left to fend for itself from an early age, unlike most plants that grow towards light, this unfortunate chap is forced to be more resourceful. With little chance of getting its feet off the ground, let alone its head in the sun, a monstera seedling's passage into adulthood is a tortuous one. At first tentatively creeping through the leaf litter, clambering over stumps, twigs and fallen branches, monsteras have a kind of inbuilt radar that sends them searching to the darkest recesses of forests until they make contact with a scalable tree. Once a juvenile monstera finds a suitable climbing frame, it wraps its clinging stems around its chosen tree and hugs it for dear life. Once established and finally able to fend for itself, it renounces its preference for shadowy places and, heaving itself upwards, it throws out whopping great leaves to make food. What the host makes of these unbidden lodgers is unclear but interestingly, some trees have ways of dissuading these unwelcome guests. Some shed their bark to prevent clinging vines and others have light-coloured bark to dissuade potential squatters from hijacking them in the first place.

And what about seeds? Back in the 1970s, sweetshop counters were crammed with shiny-egged Silly Putty, Pet Rocks, Jamboree bags, bubble gum and, most intriguing of all, Mexican jumping beans. Half a dozen brown seeds, each about the size of a pea, in a matchbox, which seemed to jump and rattle of their own accord. Magic beans, gifted by none other than the giant in the kid's fairy tale, 'Jack and the Beanstalk' – or so we were told. At the time, it never occurred to me how the old man had enough beans to supply confectioners the world over, even less that all that shake, rattle and roll is purely down to the larva of jumping bean moths nibbling their way out of captivity.

Of course, it's cheating to suggest that seeds are self-propelling, but in case you have never got to the bottom of this mystery, with the arrival of spring, female bean moths lay eggs in the fading flowers of a common shrub found growing wild on scrubby desert slopes in Mexico. Weeks later, the eggs hatch and tiny larvae worm their way inside the shrub's berries, where they feed. As the berries harden and brown, the trapped larvae get ready for one of nature's most marvellous miracles – metamorphosis. With the advent of spring rains, seeds fall to the ground, and the beans begin their lively leaping, which is thought to be an antidote to overheating as they try to jump themselves out of the sun into cooler shade. Inside, larvae begin weaving cocoons and adult moths emerge, dry their wings and fly away to take their chances in the big, wide world.

Propelled by wind, it's cheating to label tumbleweed as a major mover and shaker. But who would have thought it could be interesting? I'd always thought tumbleweed was an American native lolloping around the Wild West with all those handsome cowboys. Hah! The imposter hails from Europe and Central Asia, and took its shot at the American dream as a stowaway in a batch of seed imported from the former Russian Empire around 1870. It was an ambitious opportunist that exploited the misery of the denuded lands perpetuated by the Dust Bowl Crisis.

To say it's unpopular is an understatement. Nobody likes tumbleweed's spiny demeanour. Animals won't eat it, farmers loathe it, its pollen triggers raging hay fever and even horses give it a wide berth. Not to mention its potential to become wild fire tinder once it dries out. I'm not really selling it, am I? To give the devil it's due, the reprobate didn't voluntarily choose to leave its homeland, and tumbleweed has done exceptionally well for an unwanted migrant. Tolerant of alkaline and saline soils, wasteland or desert, it needs little water to survive, and if uprooted by wind, it boldly rolls along to its death – but not before shedding something in the region of 250,000 seeds to ensure its posterity. We're told if we have nothing nice to say, better say nothing at all, so in the spirit of making amends – tumbleweed did unwittingly help victims of the Dust Bowl catastrophe. Before morphing into thistly adulthood, its young growth is soft and pliable, driving desperate farmers to can it and sell it to supplement starvation rations as well as resourcefully cropping it for cattle fodder. In Kansas alone, enterprising migrants baled 350,000 tonnes of tumbleweed hay, an accidental contribution that helped avert the collapse of the cattle industry.

Apart from Florida, this hobo is found all over the United States. It's impervious to any kind of weed control since treating tumbleweed populations with chemicals has a negative impact on other vegetation and wildlife – once you've opened Pandora's box, it's impossible to shoehorn the sorrows back in. Still, tumbleweed deservedly owns the title of being one of the fastest plants ever to populate the United States, and say what you like, its brazen tenacity has to be admired.

But if tumbleweed doesn't impress with its persistence, you're about to meet some of the plant world's greatest pretenders that surely will.

PLANT MIMICRY: Extraordinary Adaptations

Mimicry is one of the commonest forms of human social behaviour. Subconsciously, we mirror one another's actions all the time. If someone yawns, the chances are we'll yawn shortly afterwards (watch out, though – psychologists suggest that sociopaths are immune to contagious yawning!). Plants are no less ingenious in their emulation of other species, using mimicry and/or deception, but at least humans don't have to worry about being eaten – a major preoccupation for plants, insects, fish, reptiles and mammals who play act to beat the odds. Where a harmless species physically evolves to imitate a more harmful or unpalatable species to avoid being harmed or gobbled up, thereby greatly increasing their chances of survival, is called – no doubt you've heard of it – Batesian mimicry.

It is so named for the nineteenth-century British naturalist Henry Bates, whose fascination for butterflies precipitated a journey to the Amazon, where he spent over a decade exploring its rainforests and fauna on what must have been an enthralling expedition loaded with risk. Surviving malaria, yellow fever and exposure to deadly wildlife, Bates returned to England some eleven years later, having collected over a staggering 14,000 insect specimens, half of which were unknown at the time. Thanks to his work, and long before it was scientifically proven, Bates noticed that non-venomous butterflies often resembled toxic ones, through colour, characteristics or behaviour, hoodwinking predators into believing they were every bit as threatening as the real McCoy – or they just taste plain awful. Tiger moths and gorongosa moths of Mozambique increase their odds of survival by producing ultrasound squeaks that confuse bats by either by jamming their radar or warning that they are unpalatable. Hoverflies convincingly ape wasps to avoid being devoured by predators. Peacock butterflies flash false eye-spots on their wings, pretending to be larger than they are to frighten off attackers.

And some animals go beyond mimicking appearance, resorting to vocal trickery. The amber-eyed margay or tree ocelot mimics the cries of infant tamarin monkeys, luring anxious adults within striking distance. And the oak-leaf butterfly is a master of disguise. It's nearly impossible to distinguish from a real dead leaf until you touch it – at which point the 'leaf' suddenly spreads its saffron-and-blue banded wings and flutters away.

Some agaves, aloes and palms have thorn-like impressions which look very much like the real thing but are just harmless parodies of the vicious spines inherent in other harmful species. *Corydalis hemidicentra,* clinging to scree slopes of China and Tibet, uses camouflage to blend in with the rocky outcrop around it. A small, low-growing plant, looking much like clover with flattish, blackish-brown leaves, is easily mistaken for a pebble. On one hand, its deception reduces the chances of being munched, on the other, it increases the chances of being trodden on. As usual, plants are playing the percentages.

The extent of plant mimicry is not fully understood, and behaviours vary hugely across and within plant species and are arguably attributed to physical responses or any number of factors such as cold, heat, drought and stress. Herbivores scavenging for food are heavily influenced by a plant's appearance or particular scent, so it makes perfect sense that a harmless plant might well copy the physical characteristics of a chemically defended species to its own advantage. Mimicry also helps plants extend their attraction to pollinators. A copycat plant might sensibly imitate the shape, colour and scent of another that reliably dispenses nectar rewards, except the pretender has no intention of mirroring such generosity.

We have Darwin to thank for enlightening us as to how beautiful catasetum orchids have increased their prospects by adopting a dominatrix style of seduction. Less concerned with appealing to the masses, they are minutely focused on bees. And not just any old bees, but euglossine, or orchid, bees looking for their next meal. Male orchids achieve their pollinating mission with a macho swagger by asserting exclusive breeding rights over their female counterparts via tensioned antennae found at the centre of their intriguing, lipped flowers. The mechanism instantly detects landing insects and galvanises male orchids to slap visitors with a sticky, pollen-clad disc with such force that it dislodges their punch-drunk visitors. Bees quickly learn not to linger too long and instead focus their attentions on female orchids, which offer a warmer welcome, where they can land without being throttled and obligingly slip their pollen pouches into the orchid's specially adapted grooves.

As if hand-painted by angels and astoundingly beautiful, you could easily mistake the orchid mantis for a praying one. Jointed legs and a thorax, topped with a curious triangular head precisely masquerade as petals and sepals. Orchid mantis are chameleons, able to shift colour to match their surroundings, and males and females differ in appearance too: the females are larger and flamboyantly marked, while the smaller males wear dowdier shades, a strategy that helps them ambush prey and slip past predators.

Other orchids are no less tricky. *Bulbophyllum inouei* of Borneo lacks a common name, but despite its delicate looks, this desert orchid is a born deceiver. In the harsh desert where pollinators are scarce, it's up against some pretty stiff competition, competing with flies and dung beetles – the sort of visitors lured by the stench of decay. Since it can't actually rot without committing suicide, *inouei* cheats and wafts a diluted imitation of putrid perfume. Easily fooled, beetles scuttle toward the bloom, only to be slapped with a smear of tacky, pollen-laden mucus. There's no sweet reward, of course. Duped and disgruntled, beetles wander off to find another orchid, unwittingly dropping their cargo to the next flower, and the next, falling for the same old con again and again. Like us, insects don't mind lending a hand, but they do expect to be paid for their co-operation. Orchids, however, clearly prefer the hustle.

The century plant has increased the odds of reproduction by greatly increasing the number of favourable pollinators it attracts, even though it is quite capable of reproducing both asexually and sexually. Throughout its life, the century plant produces mini-clones, guaranteeing its accession. For good measure, it still finds the strength to throw up one massive, stinking, sexually overt flower in its final days. Its malodorous meaty stench is guaranteed to lure legions of flies in addition to its usual suitors of bats, bees, moths and hummingbirds.

And only the churlish could fail to be wowed by the extravagant blooms of passion flowers, possessors of a ridiculously simple but effective deception. One of the key reasons it tends to suffer less damage from feeding caterpillars than other plants such as cabbages and nasturtiums are those rashes of small orange flecks on its leaves. You might have even mistaken them for some sort of disease or rust? Squint hard – they look almost identical to butterfly eggs. It's just cunning disinformation for female butterflies looking for a place to lay their eggs. Seeing that the foliage is already fully occupied, they are persuaded to move on and set up nurseries elsewhere.

However, a Chilean rainforest dweller, the unfathomable mimic vine has foxed all human attempts to define its falsehood over its apparent ability to shapeshift. Incredibly, it seems able to change both the shape and colour of its leaves, not only to emulate the plants it climbs but also other plants that it never comes into contact with. While agreeing the mimic vine is an exceptional plant, scientists are baffled at how it can adapt its foliage to resemble not just one, but multiple species of nearby plants at the same time. And although current research hasn't yet found the answers for this remarkable behaviour, it has opened the debate as to whether some plants have memories.

SENSITIVITY: Memory and Response

It's all very well hooting with laughter at the suggestion that plants can learn, let alone memorise things, but there is clear evidence that they can. South American shy mimosa, otherwise known as the sensitive or tickle-me plant, is a handsome individual that famously recoils at being touched. Its tactile foliage and pretty pincushion pale-pink flowers make it a popular houseplant in colder climes. If you are growing mimosa, you may have noticed it shrinks on contact, an adaption from the wild whereby it presents itself as an unappetising, sickly sort and not worth eating – a clever defence against herbivores as well as reducing water waste from drying winds. The cute corpsing routine is down to the swift release of water from its leaves and stems. Try it for yourself and see. One light touch and immediately the whole plant sags, mimicking sickness. Give it twenty minutes or so without further caress and it perks up as good as new.

Mimosa might be responding to an internal clock, because experiments in the early eighteenth century by French astronomer Jean-Jacques de Mairan found that even when deprived of light, it still moved, and he went all out to test his theory. He placed a mimosa inside a suitcase wrapped in a blanket, tucked it inside a cupboard, inside a cave, and even when deprived of all light, the mimosa still moved. I'm dubious as to how he managed to monitor its behaviour unless he was locked in the suitcase with it, but let's take his evidence on face value because modern boffins haven't exactly nailed the reasons why mimosa does what it does either.

But what I hadn't realised is that mimosa has a remarkable memory. Laugh if you like, but I was so excited at learning about its capabilities I immediately searched for supporting evidence and found numerous studies that backed up this incredible revelation. During laboratory trials (no harm was caused to the plant during experiments by the way :-), a plant was repeatedly disturbed by some special-fangled apparatus and it repeatedly collapsed in response. However, after half a dozen forced drops, the plant stopped responding, leading scientists to believe that it had *learned* – Yes! Learned! – what environmental threats it could and couldn't safely ignore. Say what you like – that's pretty captivating stuff. Especially when, while undergoing similar testing, bees forgot what they had learned within twenty-four hours, whereas mimosas retested at weekly intervals remembered the information for a full month afterwards.

Many climbing plants have a sensitive side. Sweet peas and passion flowers respond to contact with any climbing frame by sensing its proximity, curling wiry tendrils and grasping the desired support. Wild briony, another coiling climber, related to cucumber, reacts in the same way, although it suffers from a somewhat tarnished reputation by being highly toxic. Being that mandrake was fairly tricky to grow in northern climes in days of yore, unscrupulous herbalists often palmed wild briony off in its place. Given that, at best, it induces chronic diarrhoea and, at worst, can be fatal, it's probably best left to its own devices growing in the wild.

Referred to as 'the most wonderful plant in the world' by Darwin, the Venus flytrap famously snaps shut when it detects insect movement through fine sensory hairs on its leaves, and other plants have equally measured responses. Back to our fine friend, rock cress, which, when its roots encounter elements it doesn't like, simply chooses to grow away from the source of irritation – conduct that would suit us well in our human relationships!

While we're on the subject of sensitivity, there's far more to stinging nettles than meets the eye. Being highly nutritious, they top the herbivore tasting menu. You might be enjoying a barelegged stroll in the summer sunshine, but nettles have no way of knowing you're harmless. You could just as well be a hungry deer or rabbit and over the centuries they have developed an effective defence. Fine hairs on the undersides of their leaves and stems contain stinging chemicals that immediately trigger a red alert when a foreign body makes contact. Once broken, tiny hairs instantly release histamine and formic acid the moment they sense danger, compounds that cause that familiar burning sensation and painful rash.

Once you've applied the cooling sap of a dock leaf to soothe your skin, take a closer look at your stinging adversary. Nettle leaves are arranged in alternating flat pairs at right angles around the stem because, more often than not, nettles grow in shady woodland, under hedges or gloomy places where sunlight can't easily penetrate. This ingenious arrangement allows them to soak up maximum amounts of available sunshine. But let's take a look at their flowers. Or perhaps you've never really noticed they have any? Entirely forgivable, since they are pretty nondescript. But then nettles don't need flashy blooms or fragrance to attract suitors because they are wind pollinated (not to be confused with dead nettles, *Lamium*, which attracts bees). Just as ferns evolved into the strapping tribe we know today, the reproductive structures of the common nettle have been on much the same journey. Look closely and you'll see the male flowers grow at the top of the stem so they can sling their pollen on the wind from a more advantageous height than if they were growing lower down, where female flowers droop, perfectly poised to catch wind-blown pollen. (Incidentally, I read there's no cast-iron evidence that dock leaves actually work on nettle rash, it's thought they're more placebo than remedy. Maybe it's just blind faith that makes us feel they alleviate our symptoms? Anyhow, it always works for me and besides, dock leaf juice is certainly cooling as it evaporates from our skin, although it's been suggested that their leaves might contain a natural antihistamine. But like so many mysteries in the plant world, no one really knows.)

Maidenhair are undeniably beautiful houseplants with irresistible fanned foliage against fine black stems. But what divas! Deserving of their capricious reputation, for all their allure, Maidenhair have the temperament of prima donnas. Touchy when it comes to soil that is too dry or too wet, and sensitive to air that's too warm, dry or cold, only the best will do. Otherwise, they simply curl and blacken in protest. To make yours a happy home, you'll need to provide similar conditions to its homeland, where it lives the high life in cracks and rock fissures, favouring locations where there's plenty of running water, such as streams and waterfalls. You'll notice I said running water – a hint that it prefers its water to pass at a speedy lick, preventing its precious feet from standing in uncomfortable puddles for any length of time. Worn out by their caprice, I've very nearly killed mine a dozen times before I figured the best place for this darling dame is one of those self-watering pots that does the nannying for you and preferably homed in a bathroom or wet room where it can enjoy the luxury of a regular misting.

Of course, water, or lack of it, is a huge factor in plant survival and different plants have their own innovative methods of finding moisture, storing and using it.

WATER DIVINERS: Hydration

'Thousands have lived without love, none without water.' W.H. Auden, the twentieth-century British-American poet, couldn't have put it better. Water is everywhere. After oxygen, it is surely the most precious commodity on Earth. Nearly two-thirds of the planet is covered by it and incredibly, molecularly speaking, modern water flowing from your kitchen tap is the same as it was back all those centuries ago when our Stegosaurus roamed ancient forests looking for its mate. A water cycle older than time – older than the solar system that has been churning for over 4 billion years.

Staying hydrated is an automatic human instinct. Hot weather makes us thirsty, so we drink. When engaged in manual labour or intensive exercise, we continually sip water to keep hydrated. And what our bodies don't need is released in sweat. Our bodies are 60 per cent water and we are constantly using it for cellular activity. Equally, water is vital to plant survival because, being up to 90 per cent water, they need it as much if not more than we do to grow and cool off.

Yet, considering how dependent plants are on water, it's surprising just how much they waste through a process called transpiration, hanging on to less than 5 per cent of the water they absorb. In the rainforests of South America, a single tree can lose over 1,000 litres of water a day – enough to fill an average bath tub ten times over. Multiply that by the estimated 3 trillion trees on Earth and you can see that's an awful lot of sweating going on. The thing is, to make sugar for growth, every hour plants take life-threatening risks because they need to take in carbon dioxide through tiny pores in their leaves to make that energy. But when these pores are open, water is lost at a mammoth rate. For each molecule of carbon dioxide they gain, a staggering 400 water molecules are lost, so they have to get the balance just right or they risk sure death through dehydration.

Luckily, plants are expert water diviners, able to detect water in the soil and respond to it through root sensors, instinctively knowing when they need more or less at any given time. Simply put, transpiration keeps water moving through the plant, working hand in hand with another admirable process – osmosis – where water is moved via a semi-permeable mesh from cells plumped with high water concentration to areas of lower concentration. When cells are turgid and full of water it helps keep a plant rigid and upright, which is why when they are hit by water shortages or you dash off on holiday leaving your garden plants bereft, cells collapse and plants droop or die from thirst. (An excellent reason to water your garden plants either early in the morning or later in the evening when they absorb more water and tend to sweat less.)

Some plants, it has to be said, are thirstier than others. Leafy crops such as spinach and lettuce will suck up every drop offered during the growing season and gluttons like giant rhubarb need an endless sump to keep them happy. Willows are infamous guzzlers with a voracious appetite for water, dispatching their roots out in search of a good drink, although they are much maligned for undermining brick foundations in their quest to slake their thirst. If willow is planted in compatible moisture-retentive soil they have no need to conduct extensive water raids, and in dry conditions they can hardly be blamed for wanting to quench their cravings. The moral is to plant willows where water is plentiful and for Lord's sake avoid planting them near house walls. By the by, willows are easy to grow – you only have to cut a stem, poke it into the ground, and within months, it will have rooted and made a new plant.

Succulents such as desert dwellers like aloe vera and cacti are probably the first plants that come to mind when it comes to water efficiency. Specialised tissue allows them to store moisture in their leaves, roots and stems and their amazing structure doubles up as an anti-drought feature. Those fine, prickly spines are not just there to dissuade grazers or painfully needle human fingers but to reduce transpiration by collecting morning dew, which is funnelled down to their roots through those characteristic grooves.

Remarkable for its ancestry, size and water-storage capacity is the African baobab. Capable of living for up to 3,000 years and growing up to 80ft tall, it is one of the world's great curiosities. There's a great kid's cartoon depicting a zebra and a lion lounging against a barrel-chested baobab, siphoning beer from a tap on its trunk – it's well-observed humour since a single tree can store 650 litres of water per cubic metre. Considering they comprise 76 per cent water, baobabs are understandably thrifty in their consumption.

Remember those pores in leaves? Baobabs work on strict stomatal regulation. Think of it as a tree's version of human bladder control. They deploy stored water for two critical purposes: keeping the massive tree upright and providing the resources required to create new leaves at the start of the growing season. Unsurprisingly, their waistlines are often nipped when baobabs shrink during this process, but it doesn't detract from their appeal or utility. Local populations in Zimbabwe have found all kinds of innovative uses for these ancient giants – housing, bus shelters, shops and there has even been an instance of a prison. It's said that eighty-odd men could easily shelter in a baobab's wide trunk and still leave room for other guests. Whether you know it better as the bottle tree, upside down tree or tree of life, baobabs have the distinction of being a keystone species, indispensable to their local areas by creating a complete habitat. Baboons eat its fruit; birds, lizards and small insects nest in its branches; fruit bats enjoy its nectar; and clearly humans enjoy its company.

On a frivolous note, perhaps frizzy hair has persuaded you of the benefits of eye-wincingly expensive argan oil? The price is fully justified when you think that it can take up to fifty years for a mature tree to bear fruit – hence the hefty price tag. Easily living up to 200 years, these battle-hardened warriors survive blistering heat of up to 50°C in semi-deserts across North Africa because they possess powerful anti-soil-eroding roots capable of diving to depths of 35m in search of water. Their scaly trunks act as natural sunscreen and modified leaves are now vicious thorns that bear testament to a centuries-old adaption to cope with scorching conditions to reduce water loss and designed to stab grazers. Not that goats take any notice. If you've ever visited Morocco or Tunisia, it's not unusual to see foraging goats niftily perched among argan's spreading branches, proving my point that, with the constitution of metal crushers, there's little vegetation goats won't take a crack at – although rosemary is off their menu, should you be interested, even the most determined billy cocks a snook at its aromatic foliage.

It's only when we are deprived of water that we truly appreciate how critical it is to our survival and, naturally, it's vital for any number of agricultural, domestic and industrial uses. Civilisations have been built on its abundance and disappeared through its lack.

For hundreds of years, water witching or water divining was part of everyday life. Ancient civilisations including the Egyptians, Romans and Greeks had long subscribed to the technique of tapping into hidden underground springs to supply their habitations and livestock. Dowsing twigs such as hazel, peach and willow have acquired a mystical reputation over the centuries in being able to sense water. So, is it myth or magic?

In modern times, the notion that a forked stick or metal rod can locate underground springs, oil or even gold might seem absurd but even Albert Einstein believed there was something in it:

> I know very well that many scientists consider dowsing as a type of superstition. According to my conviction this is, however, unjustified. The dowsing rod is a simple instrument which shows the reaction of the human nervous system to certain factors which are unknown to us at this time.

He was citing the ideomotor effect, where human beliefs, expectations or suggestions cause unconscious muscular movement that may or may not be stimulated by our surroundings.

Dowsing is heavily enshrined in folklore, despite substantial anecdotal evidence that it works, although science has never been able to conclusively prove its efficacy. Yet even today, Californian farmers and vineyard owners still consult with local diviners. And it might surprise you to learn that some modern utility companies still use dowsing to detect leaks. I dare say you would be mightily surprised if an engineer turned up in your street with a divining rod in each hand to help him locate a leak, but although it's not a broadcast strategy, engineers working for some large corporate entities unofficially still use dowsing rods.

As impressive as their amazing talent for conserving water, plants have also come up with all kinds of clever habits to suit their unique lifestyles ...

PART THREE

LIFESTYLES: MODES AND BEHAVIOUR

How fair is a garden amid the trials and passions of existence.

– Benjamin Disraeli

Humans are one of the few species on Earth that take on challenges just for fun. We don't have an urgent need to bake, craft, jog or dive into extreme sports to survive, unlike animals and plants, who rarely get the chance to relax. Largely focused on breeding and staying alive, constantly hustling for food and shelter to raise their young, for them leisure simply isn't an option. Whether they live fast, die young or hang around for a while, different plants have their own clever tricks to handle daily challenges, operating with intrinsic efficiency.

ANNUALS AND BIENNIALS: Life in the Fast Lane

Annuals and bedding plants live life in the fast lane, glamour girls with no stomach for harsh conditions. Programmed to complete all their growing and flowering superfast in frost-prone areas, cornflowers, stocks and zinnia, to name a few, are showgirls, chorus lines that hit the stage knowing their routines backwards – germinating, sprouting, flowering, setting seed and dying in a glorious blaze of glory in under twelve months. But only if favourable conditions prevail. With only one chance in life to strut their stuff, it's in their DNA to entice every Johnny-come-lately to the stage door, so it's no surprise we love their bright, shiny faces and I always feel a little sad when their playful coquetry fades.

Driven by a desire to replicate swiftly, an annual's survival is tied in with maximising seed production. To confuse matters, we often coax exotic perennials such as cosmos, which haven't developed cold-survival strategies, to behave like annuals by choosing to grow them outside of their native comfort zones. So, an annual in one region might be a perennial in another, and to add to the confusion, we call them 'hardy annuals'. Love-in-a mist and that eternal florist's favourite, Queen Anne's lace, just aren't up to extended cold, so like annuals they pretty much die of hypothermia or frostbite in cooler regions, whereas more fortunate family members still living in the old country continue to prosper outdoors.

Making seed is all very well, however, annuals still have to think about distribution, trading off survival against longevity. Both annuals and perennials experience a kind of puberty, if you like, switching from growing greenery to flowering rapidly. Sadly, once annuals have undergone this change and thrown out their fabulous blooms, dotage and death soon follow.

Years ago, a woman passing my front garden sniffily condemned my unruly riot of brick-red pelargoniums, decrying them as common. If there's one thing I can't abide is horticultural snobbery. Any snobbery. It's as bad as someone telling you that a Hockney trumps a Warhol – completely pointless condescension. Anyway, she was obviously woefully ignorant of the fact that with such a short life expectancy, annuals live fast, love hard and die young, or she may have better appreciated their brilliance. By the way, pelargoniums aren't annuals at all. But when grown in frost-prone regions we rechristen them 'tender perennials' and treat them like annuals, although you'll find them blooming happily almost year-round in their home countries.

Not hanging around long enough to do any serious parenting, nasturtiums are natural risk-takers. Considering a single plant only produces three or four seeds per flower, it's a bit of a gamble, which explains why they flower so profusely. More blooms = more seed, taking their chances that their offspring might survive winter in earthy hideouts and sprout once warm spring weather arrives but fully accepting infant death as par for the course. For those dabbling with the idea of growing nasturtiums, in my experience, the larger the seed, the better its chances of germination, and don't forget they are great decoys planted alongside cabbages to deter the nasties. As a young gardener I couldn't be bothered with annuals, being blindsided by an endless parade of shrubs and perennials. What on Earth was I thinking? Spanish flag, a vibrant chameleon, flowering on and off all year round in its native Mediterranean, is a delight, but along with busy lizzies and tender verbenas, all are destined as sacrificial lambs that keep cool-climate gardens popping with colour all summer long.

Few of us can resist the sheer joy of annual poppies. The common corn or red poppy is forever enshrined in honouring those who lost their lives in the Great War. While serving on the Western Front, field surgeon John McCrae was moved to immortalise them in his poem 'In Flanders Fields', astonished that from the filth, gore and mud grew a beautiful red poppy. A symbol of remembrance for over 100 years, corn poppies are universally recognised as a symbol of hope and peace, and they certainly know how to go forth and multiply. With an average of over 5,000 seeds per plant, those distinctive salt 'n' pepper shakers are perfectly designed to catch the breeze, flinging themselves to the four winds like soldiers of fortune.

Often overlooked in favour of more glamorous annuals that promise instant gratification (well, almost), you'd be mad not to give biennials a go. Columbines, pillared echiums, ghosting honesty, towering hollyhocks and irresistible Canterbury bells are biennials with a different MO from annuals and perennials, going from germination to seed production over a two-year stretch. Okay, you are going to have to cultivate patience, but I can guarantee spectacular payoffs.

If annuals are the glamour girls of the summer border, biennials make meaningful guest appearances. These beauties are great for cutting gardens or plugging unsightly gaps between spring- and autumn-sown annuals, so let me persuade you they are worth your time. You might well subscribe to the view that what's the point of growing something that only grows leaves in its first year? Admittedly, lacking the showboating to get everything done and dusted within a year, biennials ration their energy. It's a routine borne out of necessity when forced emigration transports them from temperate climates to colder ones. Foxgloves, for example, usually take two years to flower, but when faced with a joyful drift of pastel-tinted spires, you'd be mad not to take a punt.

If patience isn't your thing, you can cheat. Those F1 hybrids make another appearance. If sown early enough, breeders have trained certain species to flower in year one. Okay, the leaves are a bit on the lean side but then you're sacrificing foliage for heavenly flowers. I've also recently spotted the world's first upward-facing foxglove (predictably named 'Viagra' during trials – author winces painfully), although for the life of me, I can't decide whether that's a desirable feature or not. They just look odd.

I understand that two years waiting for a flower that promptly dies might sound like a waste of time. Many gardeners dread the gaping spaces left behind, coupled with the tiresome thought of having to sow seed all over again next year. You really don't think biennials have endured two years of hard slog without thinking about their kids' futures, do you? They are prolific self-seeders – that means plenty of new plants for free. And that also means successive plants. True, they might not pop up in exactly the same spot as previously occupied but nature has a way of artfully scattering seed so they pop up in the most enviable compositions you might never have dreamt up yourself (and you get to take all the credit). If you feel the line-up isn't to your taste, just pull up unwanted seedlings.

Biennials aren't without emergency measures either. When unfavourable conditions prevail, they can speed things up, going from start to finish within one season. Given that plants can sense environmental challenges or impending danger, just the thought of a prolonged dry spell or unseasonable cold prompts young plants to fast forward and cram two seasons' growth into one. Onions and carrots are prone to throwing out premature flowers in their first year if there's been a colder than usual snap, which veggie gardeners know is called bolting. (Word to the wise here, spinach will bolt the moment you turn your back on it, so steady watering is the answer. Too little or too much will send it into a frenzy, since in its eagerness to grow, it forgets to make decent roots. Keep soil moist, not soggy, and for heaven's sake don't let it dry out or it will bolt at the drop of a hat. There – you're warned.)

Instead of beefing up their roots like perennials, biennials switch to crisis mode and divert all their energy into flower and seed production in a desperate attempt to beat the weather. Lettuce is a champion bolter, as is coriander, which is why you can buy no-bolt and slow-bolt varieties, sometimes sold as 'long standing', which coincidentally have no objection to continuous haircuts for culinary pursuits. It's a man-made device, hardly reliable but to each their own. However, from a plant's point of view, once it bolts, it's a race to the death. Throwing out flowers and seed in double-quick time prompts a bitter-tasting compound, sesquiterpene lactones, to accumulate in the leaves, a substance that increases once flowers are produced. Knowing it's do or die, bolters are damned if they're going to be eaten by insects or fail to flower when the finishing line is in sight. You have to admire their ability to swivel on a sixpence.

Naturally, you don't have to have your roots planted on dry land to demonstrate a few nifty moves, and aquatic plants offer a perfect example.

AQUATICS:
Water Babies

Sea, pond and river plants face unique challenges, spending their lives either fully or partly submerged in watery realms. It's no coincidence, then, that we find ourselves in Giverny, northern France, on an early July morning, where the morning sun is already warming the mossy waters of Claude Monet's Garden. The flowers have long faded on the wisteria. At the water's edge, a golden weeping willow dips graceful limbs into the warming shallows while the iconic emerald-green bridge spanning the pond gleams wistfully in a heat haze. Lily pads slumber in the early light and it's easy to see why the nineteenth-century French impressionist was so seduced by their charms. Sleeping peacefully under a blue heaven, the water lilies have yet to awaken. Unlike yesterday, which was clouded grey, the morning is bathed in warm sunshine and as the tourists begin to arrive, at first a trickle then building to a steady stream, visitors will have to wait until afternoon for the late-rising water lilies to rub their eyes and wake.

Unlike many sulky plants that steadfastly refuse to open under cloudy skies (poppies can be a bit temperamental that way) – it's understandable that many flowering plants want to conserve energy until the sun comes out, since sunless days means they can't make adequate energy – I suspect water lilies, like my tactile mimosa, have remarkable memories.

The non-scientific theory (according to this author anyway) is that water lilies are born optimists and will flower every day regardless. Even if the day is gloomy and wet, they seem to remember previous sunny days, prompting them to stretch and yawn, stretching in the sunlight and flattening their leaves to allow their flowers to unfold. Basking peacefully in the heat and wafting sweet floral fragrance on summer breezes, they float dreamily until dusk approaches, and then as Giverny's tourists drift away, the lilies return to their beauty sleep. Each fragile bloom lasts about a week before fading to allow fresh flowers to take their place. I can't imagine for one moment that Monet just painted these sleepy mermaids, because the warmer the weather, the more rapidly water lilies grow and are likely to take over a small pond quite quickly. I'm sure the artist didn't underestimate their appetite for living space, and I'll hazard he spent as much time trimming leaves or pulling up surplus plants to ensure they didn't swamp his mesmerising compositions, although on a practical note, you'll be pleased to hear that water lilies seldom sicken.

So, how do aquatics breathe when their roots at the bottom of muddy lakes, pools and ponds are denied vital oxygen? Incredibly, water lilies have a superb circulatory system, whereby floating lily pads pump air via their stems down to submerged roots each day, releasing oxygen and absorbing any surplus carbon dioxide before sending it back up to the leaves to be recycled during photosynthesis. Those appealing water-repellent leaves are practical as well as beautiful. Designed with tiny, waxy breathing holes to keep them dirt free, there's no danger of their becoming waterlogged. Like water lilies, India's national symbol, the lotus flower, is not only a super floater but is self-cleaning in the same way. Coming from humid, tropical regions, you might not think that water lilies could survive the chilling cold of Monet's pond in December. Fortunately, they sense winter's approach, and their defence mechanism springs into action. They slow down, their flowers fade, they stop producing leaves and slide dreamily underwater for a long winter's nap.

Diminutive water meal can't hold a light to a water lily's somnambulant charms, but it has the honour of being not just the smallest, but the fastest plant on Earth. It's related to common duckweed, tiny green balls about the size of pinheads without obvious organs, roots, stems or leaves, so at first glance, it's hard to know how water meal manages to survive at all. Approximately 40 per cent of plants have genes that distinguish between light and darkness, which limits their daily growth hours, and so most do all the heavy lifting before lunch. However, being able to reproduce rapidly in far from ideal habitats lets some species double their number in twenty-four hours, which has forced this tiny organism to slimline its gene pool down to the bare necessities. Without having to support complex structures, water meal jettisons any growth inhibitors, allowing them to motor all day and night. Spending its days and nights backstroking on the surface of marshes and freshwater lakes the world over, it also holds the modest accolade of producing the world's smallest fruit, an utricle! Behaving more like a yeast than a plant, a mother births (or buds) a daughter who, in turn, doubles in size, and within twenty-four hours and still tied to mum's aprons strings, she births children of her own and so on.

Let's not overlook a more commonplace water wader. Sturdy old bulrushes or reed mace aren't much to look at. However, in their own humble fashion, they still conduct modest miracles. They filter out potentially harmful nitrates and phosphates from the water, while their dried seed heads offer food, much-needed insulation for nest building and prove irresistible to darting dragonflies eying up potential nurseries to safely stow their nymphs.

While we tend to think of carnivorous plants living largely on land, sea anemones are seasoned mariners, albeit voracious for their pound of flesh. Would you believe those friendly, tentacled marine plants featured in *Finding Nemo* are related to jellyfish? Naturally, there's little they can do to control the flow of water around them, so they survive by getting nutrients and oxygen from local currents. Which, in turn, provides them with the energy to craftily adopt the rainbow hues of their surroundings by camouflaging themselves against rocks and reefs, while others like to submerge in mud or sand. While rocks provide safe harbour for many of our fishy friends, any unsuspecting sea life darting for cover might find themselves paralysed by an anemone's stinging tentacles before being sucked into its central mouth and digested.

Florists and keepers of exotic reptiles have long appreciated sphagnum moss, another carbon super-storer. Not exactly an aquatic plant, but accustomed to soggy habitats, including marshlands, peat bogs and wetlands, sphagnum moss, like ferns, reproduces through spores. Interestingly, it was used to bind wounds during the First World War, when scouring parties were sent out on moss-gathering forays because moss kept wounds from festering, inhibiting the growth of bacteria by keeping pH levels low. With its characteristic spongy texture, sphagnum is something of a miracle, holding over twenty times its own weight in water and preventing valuable marshlands, heaths and peat bogs from drying out – rather fortuitous when you realise that 1m of peat can take up to 1,000 years to form. Gardeners will know that peat-based compost has been on the forbidden list for many years for this very reason, and despite only occupying 3 per cent of the Earth's surface, valuable peat wetlands store a third of the world's carbon.

On our next jaunt we're diving underground again to meet a hugely dependable and varied crowd – bulbs.

BULBS: Buried Treasure

Each year, bulbs make a miraculous transformation from ugly ducklings into swans, and if you've been lucky enough to visit the Dutch tulip fields in spring, you'll appreciate just how gorgeous bulbs en masse can be. From goblet shapes to extrovert, frayed-edged chalices to cheery spring daffodils and eye-catching fists of agapanthus and alliums, their flowers are unrivalled. And rather like our earlier dinosaur, their key survival strategy relies on getting things done fast. Bulbs are changelings, emerging from earthy burrows and morphing into stunning beauties in so short a time it's a wonder they don't exhaust themselves in the process.

Simply put, bulbs are storage chambers with one short stem surrounded by fleshy leaves that contain enough energy to fuel their rapid growth. Springtime may well be bulb time, but with such a short window of opportunity, a bulb rises like a phoenix from the ashes, transformed from paper-wrapped dormancy to making roots, shoots and flowers in just a few weeks.

Although underground dwellers, bulbs aren't roots at all. Evolving in regions where there are marked seasons, true bulbs contain a complete self-build kit – leaf, stems and flower parts surrounded by fleshy scales which provide food for young plants, complete with a basal plate that produces potential roots. Slice any bulb or onion in half and you'll see all those vital components are there in tightly packed layers, all gift wrapped and ready for launch.

Bulbs are commonly lumped together with corms, rhizomes, stolons and tubers, which aren't true bulbs at all but have similar structure. Potatoes and begonias, for example, differ from real bulbs because new roots and stems grow from their eyes, something that simply doesn't happen with daffs, hyacinths and tulips. By the way, being related to lilies, its useful to know that tulips are vulnerable to the same diseases: exciting-sounding tulip fire is among the most common, a fungal pathogen that, despite its dramatic name, twists, distorts and withers. Cool, damp springs and summers are likely to set it off and the best antidote is good garden hygiene. Keep things clean and tidy, collect fallen, rotting leaves and garden debris, pull up and dispose of blighted tulips. Steer clear of adding them to the compost heap or you'll risk spreading the disease. And unlike daffodils that mass up year after year, tulips dwindle, giving us the perfect excuse to splash out on exciting new varieties every year.

Most people I know would swear bulbs are a Dutch invention. However, bulbs of old originated in Turkey, far from the Netherlands' flat, unrelenting landscape that has taken all the credit. Their distinctive-shaped blooms are a dead giveaway to their origins, coming from the word '*Tulipan*', meaning 'turban', and if you clamber among the rocky slopes of southern Turkey, you'll delight at hillsides studded with wild tulips and sky-blue agapanthus, although they have adopted a more squat design than the more desirable slender silhouettes of cultivated varieties – in the wild, it's better to be short statured and sturdy to combat the scouring salt-laden winds of the region.

Supplying roughly 60 per cent of the world's flowering bulbs and cut flowers, the Netherlands supplies bewitching blooms by the pyramid load, but perhaps lesser known is a curious moment in Dutch history. '*Tulpenwoerde*', or 'tulip madness', has been likened to an early precursor of the modern bubble phenomena when, in the seventeenth century, wealthy Dutch citizens were prepared to pay the price of a grand mansion for a single tulip bulb. In truth, it was more of a social fad, with wins and losses more in line with a casual flutter on the horses than wide-scale Dick Turpinism.

In my own experience, I know of a particular plant specialist that allegedly paid around £150,000 for three black hyacinth bulbs. With the possibility of endless propagation forever after, such rare-hued hyacinths equal black gold and you don't need me to tell you that black is bang on trend. Of course, nature doesn't really do black.

More often than not, flowers are dark maroon shades teetering on blackish-brown, but it doesn't dent our appetite for the rare or novel. And in the spirit of the Victorians before us, modern plant breeders have long been applying the best botanical brains to the inception of rare-coloured flowers – the holy grail being the elusive yellow sweet pea or true-blue rose. Covert breeding programmes the world over are devoted to trialling new and improved novelties as rival laboratories compete for lucrative sales. If they manage to pull it off, there's always a ready audience willing to pay top dollar for avant-garden curiosities.

Alliums are a sociable crowd, much preferring having their roots cramped alongside other family members than growing in isolation, so if you're short of space, they truly don't mind stepping on each other's toes and will happily grow in planters. They have proved incredibly efficient in binding erosion-prone soils with their tough, thick roots, although you'd not think of dicing and slicing architectural alliums as easily as we massacre their close relatives, onions.

Culinary bulbs such as chives, garlic, spring onions and shallots belong to the onion family, and we've been cultivating them for over 5,000 years. Barely a day goes by in our household where we don't use onions for one recipe or another and our ancestors found other uses for them as early as 3000 BC. Onion remains have been found in mummies in ancient Egypt, where women used them as pregnancy predictors. An onion was placed in a woman's vagina before bed, and if she awoke the following morning with her breath smelling of onion it was believed she was fertile or pregnant. (Arguably, not the most reliable barometer.) Greek Olympians faithfully subscribed to onion power and would rub it over their bodies, drink the juice and eat large quantities before competing, in the belief it made them stronger. Roman naturalist Pliny the Elder documented the efficacy of onions to treat just about every affliction under the sun – dysentery, insomnia, lumbago and poor eyesight.

And home gourmets may be better grounded in botanicals than they might think. I daresay you mightn't have thought that smarting eyes are triggered by an onion trying to defend itself. As the knife slices through its skin, onions sense danger and yelp for help, and if we were bats, we might hear their screams for mercy. Nonetheless, the onion feels a chef's brutal assault, its cells are damaged, and in response, it releases a bitter-tasting chemical intended to ward off sniffing animals. If you were a cat or dog, this is usually deterrent enough to leave well alone, but when it comes to a human wielding a kitchen knife, onions are forced to up the ante. A second chemical is released, propanethial-S-oxide, which dissolves in our eye moisture, resulting in a mild acid that makes our eyes stream. As I write, experts are dabbling with the notion of a no-more-tears onion. Oh dear. We're all going to hell in a handcart.

Earthy burrows are fine for some, but fearing life in the gutter, climbing plants harbour dreams of wide-open skies.

CLIMBERS: Soaring Ambitions

Wisteria is a splendid thing, an extravagant, chandeliered affair, dripping with fragrant tiers of soft lilac or white blossoms, but once established, what a relentless mistress it turns out to be. Many a gardener, vicar, lady of the manor or urbanite winces at the thought of the necessary bi-annual pruning (winter and mid-summer, should you be interested). I've pruned a 50ft wisteria that was grabbing roof tiles off a listed building to under 6ft and within three years it had reached a respectable height once more. Frankly, if you hack it at the wrong time of year by accidentally snipping off the flower buds, the worst that can happen is it throws out leafy growth with great abandon, sacrificing the current year's flower and fragrance for foliage. Wisteria cuttings can be a bit of a risky enterprise, by the way, because until they've hung around several years taking the measure of things, wisterias are a bit sniffy about blooming in early childhood. And perplexing!

It was only twenty years ago that I realised that two wisterias growing in my garden appeared to be twining in opposite directions. The more common species, Chinese wisteria, winds from left to right, but on closer inspection, its relative, Japanese wisteria, twisted in a clockwise direction. I asked several nurserymen whose knowledge was far superior to mine and only one provided a somewhat suspect interpretation that went something like this.

Apparently, centuries ago, Japan was part of the Southern Hemisphere. That's the suspect bit, since I can't find any record of Japan ever being located anywhere but the Northern Hemisphere. Still, I quite liked his story, so indulge me. It would seem climbing plants native to the Northern Hemisphere twine in an anti-clockwise direction whereas those growing in the Southern Hemisphere wind clockwise – '*It's down to gravity, Missus!*' – likening it to water disappearing down a plug hole. More likely, it's an environmental adaptation, but if there's anyone out there who knows the truth of the matter, I'd be mighty glad to hear it.

So, why do climbers climb or flowering vines and beautiful rambling roses know which way is up? Stems and leaves are working independently on negative gravity, whereas the roots are responding to positive gravity. That might seem enough to ensure their one-upmanship in a lush garden where competition for food and water is more genteel, but climbers have a few more refinements that ensure their daily survival. Rose thorns are actually modified leaves that allow climbing and rambling roses to heave themselves up above the opposition, and the sharp thorns also help dissuade notorious nibblers like muntjac and rabbits, although as I've mentioned, goats aren't put off by a crunchy snack. For those disinclined to don gauntlets when pruning to protect against thorns, vibrant pink climbing *R.* 'Zephirine Drouhin' and cloudy apricot *R.* 'Kew Gardens' are *almost* thornless (since the absolutely thornless rose, like a yellow sweet pea, has yet to be bred). Unable to support themselves gracefully, roses aren't naturally equipped to climb solo, which is why they piggy-back on their neighbours using their thorns as grappling hooks. In case you were unaware, roses are repeatedly voted the world's favourite flower. It's no small wonder – few flowering species offer such diverse size, shape, perfume and colour. Although roses don't do biblical monsoons or hailstones very well, most can cope with drought, and many repeat flowerers will send out shivering blooms as late as December.

Outside my window there's an old tree so camouflaged in ivy, honeysuckle and fragrant Paul's Himalayan musk that, for the life of me, I couldn't tell you what type of tree it is. A veritable apartment block of hustle and bustle, chiming with gossip and bickering blackbirds, chaffinches and song thrushes. Rambling roses can grow up to 40ft in the right place (another shrub that thinks itself a tree) and being superbly equipped for ascent, it easily elevates itself above the rabble. Once they've broken through any lower undergrowth, there's nothing to impede their soaring ambitions.

After flowers come seeds. In this way, most flowering plants ensure future generations and each time we cut fresh blooms to bring indoors, in effect, we're depriving them of sending forth their progeny. A pricked finger as we cut a thorny rose for the vase is nature's way of reminding us of its daily struggle for life.

Jasmine benefits from a bit of scaffolding, otherwise it quickly becomes an unattractive tangle, while other climbers have found equally effective ways of keeping one step ahead. Ivy, especially common or English ivy, is happy to trounce any challenger, growing in all directions, sideways, up, over and under. It's said that ivies tip-toe in their first year, walk in their second and gallop in their third – and when it comes to English and Irish ivies, you'd better believe it. Specialised aerial roots allow ivy to insinuate itself into any unlikely foothold, under tiles, in brick mortar or in fissures in tree bark, all reliable launch pads to indulge its greater aspirations, hoisting it skyward toward sunlight. (If you want some degree of control, choose one of the more compact types, there are some excellent choices out there.)

Virginia creeper, on the other hand, has adhesive pads that allow it to cling to almost any surface, whereas clematis, cucumbers, grape vines, melons, passion flowers and sweet peas depend on strong, wiry tendrils. You'll probably not notice how tough they are until it comes to pruning them away from a neighbouring shrub or tree in which they have become entangled. Determined snipping and patient surgery is required to free them as they corkscrew around neighbouring stems in an effort to overreach their competitors. Born vertically challenged, it's surprising they turn out to be such able climbers. In infancy, they are forced to stumble around at ground level until they meet something they can climb – a bamboo cane, trellis or the base of a shrub, anything they can grab hold of, because they possess unique intuition and, like the Swiss cheese plant, experience a chemical change that lets them suddenly defy the lateral gravity of their birth. Once firmly ensconced, they weave their way upward, happy and unbridled, with their heads in the sun and their feet in the shade.

And while we're on the subject. Bindweed. A rampant, perennial twiner, engulfing everything in its path, yet its annual cousin, morning glory, is much admired, unless of course you live in warmer, non-frosted areas, when you might regard its flared purple trumpets an equal nuisance. Both are quick-growing flowering vines, easily reaching 15ft in just a matter of weeks. It's easy to tell the difference: bindweed has arrow-shaped foliage and morning glory has rather more pleasing heart-shaped leaves. Neither can cope with frost but bindweed is an exemplary survivor and will self-sow with gay abandon. Given favourable conditions, seeds scattered from a previous year dependably regrow into equally determined offspring and you don't need me to tell you that bindweed is an absolute sod to get rid of. Roots can spread several feet underground, so hand weeding is a complete waste of effort. Driving bamboo canes into the ground is the best way to be rid of it. Hung by its own petard, that innate clambering instinct takes over and once it has coiled around a cane, just take a glyphosate-based weedkiller and stroke it over the leaves, which carries the poison down to its roots and purges the nuisance once and for all.

Mind you, we shouldn't tar all bindweeds with the same brush. Silverbush is an altogether more refined family member, a stunning compact pewtered shrublet, rarely reaching more than 60cm in height and studded with glistening white parasols in summer. Pure bliss. Field bindweed, a surprisingly pretty pink and/or white vining species commonly spotted creeping along the ground in fields or sandy wastelands, can live in arid desert but, even then, I suggest it needs at least *some* water to thrive.

Even desert plants need a drop of moisture now and again, and as we're about to find out, they have cleverly modified their growth patterns to conquer one of the most perilous landscapes in the world.

DESERT DWELLERS: Cacti And Ephemerals

Deserts. Harsh dry landscapes where only the tough survive. Rainfall and food are scarce. A blistering habitat with a famine of the key element plants need to grow – water. And even with occasional rainfall, sand is a meagre diet – it won't keep a plant nourished for long.

With the misfortune of being born in one the most perilous habitats on Earth, desert plants are more interested in reproduction than long-term survival, not through preference but necessity. Desert blooming is an inexact science and only those plants with entrepreneurial instincts can hope to flourish. Seeds take the risk of burying their heads in the sand waiting until fate intervenes, and as we've seen, they can remain viable for years just waiting for the right conditions to germinate, although some seed coats contain a growth inhibitor that can only be washed off by rain.

And in deserts any old downpour simply won't do. Desert seed doesn't need lashings of rain: too much washes them away; too little and it doesn't penetrate the ground deeply enough to be of any real use. They need the *right* rain, in the *right* amounts, at the *right* time to fulfil their destiny. You only have to glimpse a scorching tide of red poppies in California's Antelope Valley in spring to appreciate a fleeting moment in time or the sudden bloom of wildflowers in the Mojave Desert that allows us to witness one of the most spectacular, ephemeral landscapes on Earth.

From the deserts and arid mountains of the Americas to the extremes of Patagonia, cacti are champion dune warriors, facing untold hardships without flinching due to their ability to store water, coupled with a fleeting growing season. Their strategy is straightforward: once desperately needed rainfall occurs, cacti absorb water through their roots and store it in swollen stems. Their roots have adapted to spread wide and deep to maximise water capture, also aided by leaves modified to spines to prevent them from being eaten by opportunistic grazers, and that handy waxy coating helps minimise water loss.

Either loved for their sculptural shapes and surprising flowers or loathed for their prickly demeanour, cacti know they have only one chance to make a first impression. The underdogs in dry, arid landscapes, cacti might throw out a single or a few flowers in a season. Some are sporadic, with blooms lasting just a couple of days, while others never flower at all. Although they flower infrequently, they more than make up for it by producing plentiful seed, since they know they are only going to get a limited crack at making offspring.

Relying intuitively on night-time pollinators such as haw moths and bats, mysterious night-blooming cacti have intelligently learned to synchronise their behaviour to accommodate their nocturnal coupling. But being prickly isn't enough to see them through relentless hardship, so over time they have developed a very different metabolism from most other plants. If cacti behaved in the same way as, say, herbaceous plants, respiring and photosynthesising during daylight hours with their pores wide open, the engine room would be put under fatal strain. Enter a life-saving device, CAM, or crassulacean acid metabolism, that allows their engines to idle by putting photosynthesis on hold, effectively isolating plants from the outdoor environment. By closing their stomata during daylight hours, they conserve energy and water and cleverly delay the opening of leaf pores until nightfall when temperatures are lower and humidity levels are higher – a hyper-effective process compared to normal respiratory water loss and up to ten times more effective. Cacti are still drawing in carbon dioxide, which is stored as malic acid until daylight arrives again, when – ta-da! – it's magically turned back into carbon dioxide and, with sunlight, is then transformed into sugar. Pretty marvellous conjuring, no?

If you happen to come across a scaly dried tennis ball at your local garden centre, you might think the staff have been rather slack about watering their stock. With the ability to rise from the dead, the aptly named resurrection plant is something of a novelty, and a remarkable example of resistance and renewal. A doughty, rootless dweller from the deserts of Central America and Mexico, it can lose up to 90 per cent of its cellular water and curiously plays dead during extended drought. In truth, it doesn't so much pass away as enter a suspended animation, reviving as good as new once rainfall returns. Happily, a light shower is all that's needed to transform this scruffy-looking individual into handsome rosettes of emerald green within twenty-four hours. And each time it's hit by extreme aridity, it pulls off its Lazarus routine, time and time again, allowing it to live to a ripe old age.

Alligator plants are no less inventive in their determination to foster their bloodline. Otherwise known as the mother of thousands, there's no better description for this incredibly attractive desert dweller that likes to keep its brood close. It cossets miniature plants resembling rows of tiny teeth that fringe its leaves, hundreds at a time – each plantlet a miniature replica of mum. At any time, baby plants simply drop off and, fully equipped with roots and leaves from birth, they have all the kit they need to survive independently. With fleshy leaves arranged in slanting circular rosettes, mother of thousands captures and stores any rainfall, easily funnelling it down to its roots, while the waxy sheen helps reduce water loss and it can store enough water to keep motoring even when there has been no rainfall for months.

Desert or Tortula moss behaves in similar fashion. But why should such a drab little moss be of interest? Because after being locked in a freezer at –80°C for five years, and exhibiting an extraordinary resistance to gamma radiation, it suddenly finds itself the centre of attention. Recognised as one of the most stress-tolerant organisms on Earth, this self-effacing individual is being sized up for the daunting task of possibly populating extra-terrestrial environments. Even after losing approximately 98 per cent of its water content, it rehydrates within seconds with its physiological and photosynthetic abilities immediately intact.

It was over 100 years ago that British archaeologist Howard Carter discovered the tomb of Egyptian King Tutankhamun, the only Egyptian tomb found completely untouched. Overshadowed by that famous gleaming gold death mask, on his chest lay a humbler jewel – a yellow gemstone, shaped in an intricately carved scarab. Ancient Egyptians believed dung beetles to be divine, a representation of the cycle of life, birth, death and regeneration, and who'd argue that a dung beetle might prove a meaningful companion if you are about to embark on a journey into the afterlife? Numerous jewelled scarabs have been found in ancient burial chambers of the rich and powerful, for it was believed that dung beetles mirrored the Sun God, Ra, rolling the sun over the horizon and bathing the world in daylight – only scarabs improvise with dung balls.

It's often forgotten that dung beetles are industrious recyclers, playing an immense role in improving semi-arid habitats by enriching the soil, and they are found on just about every continent except Antarctica. A bit like rag and bone men, who collect any old junk from city streets, you'll have spotted these busy little chaps on nature programmes, rolling balls of excrement underground to feed their larvae. Superb detectors of moist manure from camels, cattle, goats, elephants and kangaroos, dung beetles fly several miles a day in search of fresh dung pats, using specialised antennae that allow them to locate ripe pats from the air. Hoovering up debris and excreting their own waste nourishes the soil and this seemingly straightforward housekeeping is as impressive as their role in pest control. Bush flies have long been a serious scourge in Australia with an estimated 12 million cattle each producing on average around twelve cow pats per head and attracting 3,000 fat flies daily – totalling billions every year. Like everything in nature, there's a reason and a season, and from the 1970s onwards, European dung beetles have offered unrivalled 100 per cent eco-friendly pest control, spearheading the reduction of nuisance fly populations. Flies lay their eggs in pats, unaware their efforts are wasted since foraging beetles suck out all the available liquid, and with no moisture available, fly maggots simply shrivel and die.

And how many times have you cut an apple in half to find a maggoty individual lurking within? Maggots are like lepers, shunned by gentle society with an unfortunate propensity for turning up where they are least wanted. That said, we all have to start somewhere, and grow-your-own enthusiasts will know of the cunning needed to outwit unwelcome visitors to the produce patch. But to be honest, looking at a plump peach or lush raspberry, it's hard to believe that fruit and vegetables possess any survival strategies at all.

EDIBLES:
Fruit and Vegetables

Sweet, juicy berries, rosy apples, earthy beetroot and a plate of crisp French fries all look hugely enticing, but surely, as we've seen with other plant behaviours – the very opposite should be true? Make something deliberately unattractive and you can be sure it will be left well alone. However, Mother Nature had other ideas and designed fruit and veg to attract as much attention as possible.

Their coping strategies have evolved in their lands of origin and hold true even when they are transplanted thousands of miles from home. Stoned fruit, as we've already explored, have no way of releasing their seed easily. Given the choice between hanging around, over-ripening or hoping for the best, nature endowed them with a more convenient way of spreading their offspring. You can bet any produce that looks enticing to us, also looks pretty tempting to wildlife, and in their own inimitable fashion, their very attractiveness ensures their lineage. Making fruit and veg look appetising invites birds and scavenging wildlife to feast, ensuring would-be heirs are passed through animal faeces.

If fruit trees happen to live in isolated spots, making it difficult to attract plentiful do-gooders, their fleshy scions simply rot and fall to the ground, releasing the pit that, with luck, lands in a suitable spot to set up home. Mulberries, for instance, lack the ability of ripening all at once. They ripen over several weeks, playing the odds. If the weather suddenly turns nasty, they should have flung out enough decent berries to ensure their survival, and if good weather persists, they have every chance of attracting as many hungry birds as possible in round two.

It goes without saying that produce comes from a diverse range of regions and belongs to different plant families, yet it is surprising how many can be easily grown far from their native lands. Temperature is a biggie. Warm-season fruit and veg, such as bananas, eggplant, melons, peppers and tomatoes, need a longer growing season and higher temperatures to ripen. It's virtually impossible to get bananas to ripen outdoors in northern Europe; summer temperatures simply aren't high, humid or long enough. Yet colder regions can still achieve fat, tasty aubergines, figs and tomatoes in a decent summer, although admittedly, cantaloupe, grapes and watermelons are a bit of a struggle unless you can offer hothouse luxury.

Historians and modern scientists hotly contest the origins of maize, but at least its geographical location is widely agreed. It was several thousand years ago that the ancient Mayans discovered corn. Its ancient ancestor, a wild grass called teosinte, was a coarse plant with half a dozen unappetising green kernels stacked one on top of the other and looked nothing like today's plump, golden cobs. *Teosinte* was the lifeblood of the Maya: it was dried to make flour or soaked to make bread, tortillas and tamales and featured heavily in their religious culture. Time passed. Mexican maize spread to Peru, and when Christopher Columbus discovered America, he also discovered corn along with chilli peppers, peanuts, pineapples and potatoes. Not that anyone in England knew what to do with this new crop when he first returned to English shores waving it aloft.

I daresay no one was more surprised than corn to find itself suddenly thrust into a cold climate, wondering how on Earth it was to cope in such odd terrain and still blithely oblivious to its glowing future. Because in the 1950s, a botany professor from the University of Illinois, one John Laughnan, was busily investigating corn's gene pool when he discovered that a particular gene in the inner kernels stored less starch, which caused harvested cobs to shrivel. Fortuitously, he also noticed that it contained up to ten times more sugar than ordinary field corn and tried flogging the idea of an improved cob to food producers. The American sweet corn industry was sceptical of his findings, but notwithstanding, Laughnan determinedly carried on inbreeding this super sweetness into the prevailing varieties of the time and was the instigator of the deliciously sweet cobs we know today.

But where would a corn cob be without a grind of black pepper or a pat of creamy butter? And shame on the chef who serves spaghetti without garlic or *insalada tricolore* without torn basil. We're heading off to Scarborough Fair on our next leg, in search of parsley, sage, rosemary and thyme.

HERBS:
Healing and Legacy

Man has long relied on plants to provide relief for all kinds of conditions because our herby friends contain all manner of potent biochemical compounds. As early as 3000 BC, ancient Chinese records show the many uses of medicinal plants. African and Native Americans have relied on herbal therapies for centuries and research shows that no matter where people were located across the world, they were all using much the same plants for the same healing purposes. Ginseng, cinnamon bark, fig, juniper and willow are among the many hundreds of plant-based tinctures, teas, balms and balsams used by our forebears that are still actively used today.

It was in the early nineteenth century, when chemical analysis was still in its infancy, that apothecaries began to extract active plant ingredients and experiment with them, and in modern times, approximately a quarter of pharmaceutical drugs are still derived from botanicals. Herbs have a rich legacy in human history, serving as essentials in our daily lives.

The Greeks and Romans fashioned head gear from the aromatic bay tree and wore them as symbols of power and triumph. Greek Olympians were honoured with laurel head wreaths, Julius Caesar sported his to demonstrate his power, whereas the Emperor Tiberius adopted his crown of greenery to protect against being struck by lightning. From status symbols to treating ills or flavouring our food, these remarkable plants have played a hand in healing, nourishment and even warfare. As far back as 74 BC, the resourceful King Mithridates fought against Roman domination using the lethal power of nature. He concocted a potent blend of nectar and pollen to create 'mad honey' – a fatal substance that he used to poison advancing enemy troops.

One of the most fascinating treasures from the past is the Ebers Papyrus, a 3,500-year-old Egyptian manuscript that serves as an extraordinary archive of ancient healing practices. Imagine leafing through its pages to discover treatments like garlic for healing wounds, stinging nettles for alleviating aching joints, and dandelion, prized for its ability to combat colds and gallstones. By the time the medieval period came around, herbs were ground into powders, tinctures, teas, ointments and poultices, becoming the go-to cures for numerous ailments. Sage and mint are still effective in settling upset stomachs, while lavender continues to calm, soothe headaches and promote sleep as it did centuries ago.

A favourite myth buster is the speculation on God speaking to Moses on Mount Horeb, which has given rise to one of the most quoted passages in the Bible: 'The angel of the Lord appeared unto him in a blazing fire amidst a bush ... and behold the bush burned with fire but was not consumed.' All kinds of explanations have been offered to explain the bizarre incident. Perhaps Moses was hallucinating? Or was he simply overawed by God speaking to him from a blazing shrub?

There is, though, a perfectly logical explanation. Any plant containing botanical oils is usually flammable. Lavender and rosemary famously contain flammable oils, although the most likely biblical culprit is dittany, aka the gas plant or burning bush. A herby member of the citrus family with a surprising potential for self-combustion, dittany is coated in highly flammable oil that, when temperatures soar, has been known to burst into a flash flame before fizzling out.

Unharmed and unperturbed, it grows happily on, sadly leaving you with singed eyebrows.

Whereas blue-eyed borage keeps its cool through having hairy leaves, its foliage reflecting light rather than absorbing too much of it, and was commonly used in medieval times for chest infections and as a sedative for calming seizures. You might have noticed its similarity to viper's bugloss, a biennial herb traditionally used as an antidote to snake bites and scorpion stings.

But while being healing in the right hands, some plants have more sinister aspects to their personalities. I can't help wondering whether Shakespeare was a little obsessed with their toxic nature and suspect the bard spent a good deal of his time researching poisons for his plots. As the three witches of *Macbeth* seethed about their cauldron cackling with foreboding, it surely demonstrates that herbal potions and their possibilities were deeply ingrained in Shakespearian culture – 'Eye of newt and toe of frog, Wool of bat and tongue of dog ...'

At first glance, the ingredients for this formidable spell seem to comprise largely of reptile and animal remnants, although I'd wager a more probable explanation is that some of the additives were more likely herbs and wildflowers. 'Tongue of dog' is presumably hound's tongue, related to borage and heavily toxic, with clinging yellow burrs that snag on just about anything from sheep to human clothing, and poisonous buttercups were known as 'frog's toes'.

Henbane or yew were the most likely suspects in the murder of Hamlet's father – both fairly fast acting and fatal in the right doses. His star-crossed lovers Romeo and Juliet proved the architects of their own tragedy when Juliet faked her death by swallowing an unknown poison. Carl Linnaeus himself, the father of plant taxonomy, first catalogued the evils of *Atropa bella donna*, aka devil's berries or deadly nightshade, a poisonous perennial and most likely the poison taken by the unfortunate Juliet. Belonging to a diverse family of plants, including potatoes, whose leafy crowns are toxic – which accounts for all that tedious earthing up – neatly links this murderous relative to starchy spuds.

That said, you'll be relieved to learn there are some more well-meaning plants populating the family tree, such as chillies, peppers and aubergines. Interestingly, bella donna (meaning 'beautiful woman' in Italian) was the height of fashion in Venetian times when renaissance women found cosmetic uses for nightshade berries, making eye drops to enlarge their pupils for a 'come hither' look or powdering berries to rouge their cheeks.

Probably one of the best-known herbal references is that of the sixteenth-century British herbalist Nicholas Culpeper, who apparently spent his teenage years at Cambridge University, ostensibly to study theology, although it seems he spent as much time drinking and smoking – the more things change, the more they remain the same! After a series of career setbacks, and unqualified in any real medical sense save for a few local plant hunting expeditions, he established his own practice in Spitalfields, London, away from the prying eyes of the Royal College of Physicians, who suspected him of quackery. Dispensing cheap cures and affordable treatments to increasingly loyal patients prompted Culpeper to set down no fewer than 369 herbal medicines in his hefty tome, *The English Physician*, better known as *The Complete Herbal*. At the time, he can't have imagined that it would become a consistent bestseller and to date has seen over 100 editions.

You may have noticed herbs are often the last plants to sicken. Aromatics differ from the floral perfume of other flowering plants as they are not there for *our* pleasure but for protection. Nasturtiums always get riddled with holes from caterpillars, but planted alongside sage, oregano and basil, trust me, your herbs will remain in pristine condition. It's rare to see mint chewed by caterpillars because aromatic foliage is the very feature that safeguards herbs against predatory pests and disease. While to many, the Sunday roast is unthinkable without mint sauce, animals can't bear the taste of herbal oils, although cats, being contrary creatures, often chew grass or nibble catmint – invariably followed by vomiting hairballs (usually on the carpet, ignoring the handy wipe-clean hard floor a few inches away). One must assume it has some sort of cleansing effect, although a herbalist friend tells me that catmint has an aphrodisiacal effect on our furry friends, which might explain why they roll around in catmint borders and flatten the lot.

Aromatic silver-and-grey-leafed herbs, such as thyme, oregano, rosemary and sage, are accustomed to long, hot Mediterranean summers and have handily sun-screened themselves into survival. Those high temperatures we so relish can be a bit of a nightmare if you're a fair-skinned plant, and that appealing glaucous colouring is a defensive cuticle that deflects sunlight and avoids scorch. And if you look at the way leaves are arranged on silvered herbs, you'll see they are placed precisely to shade the plant as much as possible to prevent sunburn.

Like parsley, mint can thrive in light shade, whereas basil appreciates sunshine and warmth in cooler climes. In scorching climates basil does far better in gauzy shade, although it's a champion bolter. Once you see those tiny white rosettes appear, pinch them out to ensure you can keep harvesting its leaves for all those delicious pasta dishes. As popular for their edible uses as practical ones, nowadays we're just as likely to freeze blue borage blooms in ice cube trays to liven up cocktail hour. Lavender ice cream is always a crowd pleaser, and we sprinkle flamed nasturtiums and sweet cicely on our summer salads.

Shifting away from herbs, we're off to meet some stalwarts stranded far from home, who put up with less than ideal conditions to indulge our craving for smart interiors, yet still manage to thrive.

HOUSEPLANTS: Plants Out of Place

It's estimated we spend up to 85 per cent of our lives indoors and even as we're bent over keyboards or frowning at computer and TV screens, numerous studies suggest indoor plants go on caring for us, even if we're unaware of it. On a modest scale, houseplants improve air quality, reduce airborne pollutants and replace carbon dioxide with fresh oxygen while increasing ambient humidity, so they have proved especially beneficial in offices and workplaces, boosting our creativity and increasing our productivity by as much as 15 per cent. As popular in providing trendy interior decor, did you ever look beyond their appealing foliage and wondered where they truly belong?

Houseplants are simply plants out of place – yet, left to their own devices in their native homelands they can attain impressive stature. So, how does a plant growing naturally in a humid, exotic location survive equally well on a chilly British or North American windowsill?

Novelty houseplants, including popular pebble-like succulents such as living stones, thrive on dearth and rarely need watering. I'd hazard most house plant fatalities are down to hapless carers who either overwater, underwater, let them sizzle on sun-pooled coffee tables or freeze on chilly windowsills. Arranged attractively on a sideboard, believe me, indoor plants are way out of their comfort zones. Being native to areas of South Africa, where rainfall is scarce, living stones rely on mist and fog as their main source of moisture, so it's easy to see why they appreciate your home comforts. Light, warmth and an occasional misting if they are lucky tricks them into believing they have never left the sandy slopes and gravelled ridges of their homelands.

Similarly, our hallways and conservatories are adorned with plants that typically live in or under jungle and forest canopies. How often do you have to water a spider plant in the bathroom? Probably infrequently, since they take all the moisture they need out of the air. And cheese plants wallowing in domestic quietude in smart city terraces rarely pine for their homeland rainforests of central America, where they are decidedly less well-mannered residents.

A cheese plant or monstera on home territory makes a giant vine, ranging up to 70ft tall, but finds it impossible to survive without help. It manages to keep body and soul together by the use of aerial roots climbing through branches of neighbouring trees and producing flowers and fruit. But why that curious punch-holed foliage? Growing close to the forest floor where the sun barely penetrates, young monsteras lack the perforations of maturity. It's only as the vine matures and begins its ascent that those trademark cut-outs develop. You can imagine a vine in full leaf carries a hefty weight, and while a host provides reasonably reliable support, it can't do any harm having a back-up plan – a wind-filtering strategy. Holes in the foliage allows wind to pass through and let light filter down to the juvenile foliage below. Similarly, oak leaves are larger at the bottom than higher up for the same reason. And to be on the safe side, leaves are arranged in a fan shape around the stems to avoid any larger leaves higher up shading those growing lower down. In case you're in any doubt, monstera leaves know precisely just how many perforations are required so that energy production isn't compromised. Ingeniously, several pre-programmed suicide cells ensure that the right number of perforations die off, leaving the surrounding cells unaffected.

By now, our cheese plant has defied the odds, and its mind turns to family planning. Producing flowers and then large fruit that takes over a year to mature, its distinctive offspring are those warty, yellow breadfruit found at your local grocer's. To protect its youngsters, breadfruit ripens from the bottom up, with hexagonal, armoured scales falling away as the fruit matures. If a scavenger attempts to eat the fruit before it's ripe, a harmful substance secreted through the skin causes severe stomach, mouth and throat irritations that soon warns them off. That benign plant sitting on your coffee table has certainly got all its bases covered. On a more practical note, if you notice your plant starts weeping from time to time, it's more likely sweating than sobbing. Water droplets on its leaves are probably a sign you're overwatering, causing plants to expel excess water, so it's probably time to revise the watering routine.

Elephant's ears, with those lush, striking shields, naturally thrive in tropical and subtropical zones from Asia to Australia and are hugely popular indoor plants. Quite commonly, you'll spot them lounging languidly in coastal towns across westerly and southern parts of the British Isles in far from perfect environmental conditions or squeezing their feet into pots that never seem to get upsized. Restricted roots will always hinder a plant's potential size, which explains why house plants rarely reach the giddy heights of those growing wild in their natural habitats.

Poinsettias offer a perfect example. Native to Central America, and more specifically, south Mexico, where the ancient Aztecs traditionally used their vibrant dye for clothing, Christmas flowers flourish where summers are hot, long, overcast and humid and winters are short and mild with temperatures rarely falling below 63°F or rising above 98°F. Visit any supermarket or garden centre over Christmas and they are awash with potted poinsettias with their eye-catching Chelsea pensioner scarlet blazers that look so perfect you could easily be forgiven for thinking them artificial. Eternally popular as festive gifts destined for a favourite relative or a sunny windowsill, once poinsettias have ceased to thrill, they usually die from neglect or are discarded once their Yuletide showboating is over.

Quite by chance a couple of years ago, a vibrant crimson cloud flowering by a dusty roadside in the eastern Mediterranean made me pull over for a closer look. There, to my surprise, bearing little resemblance to the dwarfed supermarket offerings back home, studded with vivid red bracts and easily twice my height, was a festive poinsettia flowering in all its natural glory. Naturally, poinsettias will never achieve such extravagance in colder climes and are much safer consigned to a warm conservatory or frost-free porch, but it demonstrates how plants adapt to hostile environments and still bloom and delight in less-than-ideal conditions. To paraphrase, a poinsettia is for life – not just for Christmas.

Probably the best-known plants are perennials. Our gardens and parks are crammed with them, and they do exactly what it says on the tin – recur and endure for years. Perennials are superbly adapted to their lifestyle and have so firmly ensconced themselves in our affections that we're happy to lend them a helping hand at every opportunity.

PERENNIALS:
Long-Lived Plants

Among plants, ornamental perennials are highly favoured. They get the best attention, the best soil and, like dogs at Crufts, excessive pampering and grooming. Non-woody perennials (or most soft-stemmed garden plants, for want of a better description) have worked out that they simply can't survive a winter freeze, which causes fatal cellular damage. By disappearing underground and remaining dormant through the harshest months they avoid damage to tender tissues, with the soil providing insulation for those all-important roots. A protected winter slumber is all that's required until they sense warmer weather approaching, prompting new stems to grow from the crowns each year. But pop up too early and the leaves or early flowers get frostbite, so their body clocks save them from impending disaster.

In reality, perennials aren't detecting temperature change at all. They have a secret indicator, something called phytochrome, a molecule that reacts to nights getting shorter and daylight hours growing longer, allowing them a trouble-free transition from winter to spring. Once they detect rising temperatures, they know it's safe to stick their heads above the parapet once more.

Of course, some perennials live longer than others, ranging from a few years to a few decades. It's said if you plant a peony, you'll have it for life. My late mother-in-law had a fine stand of *P.* 'Sarah Bernhardt' that she swore has been flowering its socks off for nigh on forty years, and since they can last 100 years plus, hers was, relatively speaking, still only middle-aged. Depending on where you live, peonies mean different things. In China and Japan, their blooms symbolise abundance, honour and wealth, while Pliny the Elder exhorted them for other reasons entirely. Both red and black seeds were used to delay menstruation, treat bowel and bladder disorders and, not least, prevent nightmares. However, caution was advised against picking them in the woods at night – attracting the wrath of the woodpecker of Mars from Roman mythology when any furtive filcher would suffer anal prolapse as it pecked at the thief's eyes!

Columbines and gorgeous self-supporting purple vervain are often referred to as short-lived but they self-seed so prolifically that you'll barely notice they've passed on. Admittedly, roses can get bare and leggy lower down after a number of years, as does lavender, but contrary to books telling you it has a short shelf life, lavender can bloom for years, albeit never quite recapturing the plumpness of youth. Mediterranean plants have a reputation for being short-lived, but I suspect that those growing in colder regions tend to have a shorter life expectancy than those growing in their more clement homelands.

A quick mention for vernals, better described as fleeting spring wonders that grab the opportunity to shine early in the year. The earliest of bloomers, vernal perennials are largely a short-statured bunch, lacking the time to attain any reasonable height and able to produce shoots, leaves and flowers in a short space of time. Crocus, cyclamen, intricately flowered Dutchman's breeches, trout lilies, trillium, early iris and snowdrops are all fleeting bloomers, disappearing beneath the soil when other perennials are just getting going but providing a vital food source for early bees in the process.

In colder regions we feverishly wrap exotics such as bananas, cannas and more vulnerable soft-stemmed plants in fleece with the onset of winter because the cells that transport water from the roots to leaves could prove fatal if left to freeze. Yet, we seldom afford the same cosseting to shrubs and trees, forgetting they are also perennials – but woody ones.

Rugged up in thermal outerwear, hardy trees and large shrubs can forgo such cosseting. They navigate winter perils because they simply stop growing. Well, not quite. They may look if they are doing nothing, but already next year's leaves are being formed, which is pretty miraculous when you think that their only annual vacation is more of a busman's holiday. Having shed their leaves, deciduous trees and shrubs have no need to photosynthesise and effectively shut down their internal processes, relying on their woody coating as insulation to protect the delicate vascular tissues underneath.

Conversely, conifers and evergreens retain their foliage and run the gauntlet of harsh winters. They still make energy, but it comes at a price.

High in the rugged snow-capped mountains, a solitary spruce stands cloaked in crystal silence. Its velvet green needles bend under the weight of freshly fallen snow, its boughs quiver with the effort of holding strong against chill roaring winds streaking through an alpine pass. Frost clings to its bark and in the pale light of a waning sun, all around, the world is swaddled in a white hush.

Admittedly, an enticing landscape for a walk through a winter wonderland, but below-freezing conditions, heavy snowfall and storms are a significant challenge for coniferous trees. Without their dense greenery, conifers can't photosynthesise and so they have no choice but to brazen it out. Since all trees vary in their cold tolerance, a winter wardrobe is essential.

Luckily, years of adaption to their local climate equips them with the necessary tools for the job. Water, as we know, freezes at

32°F and that same expansion happens in evergreens. It's a chemical anti-freeze, if you like, that also accounts for their aromatic scent, a coping mechanism that reduces the amount of water held in the leaves, together with a tough, waxy coating that effectively stops their leaves from freezing.

That might avert the problem of damaged leaves but what happens to the significant amounts of water held in the branches and trunk? Ingeniously, conifers control when and where ice forms at a cellular level. By moving water from inside their cells to the spaces in between ensures that if the water freezes it doesn't impair their functionality.

Although evergreen trees and shrubs are better at combatting adverse cold temperatures than those that lose their leaves in winter, their fully clothed branches act like sails on a boat. Under the onslaught of ferocious winds, they suffer branch breakages that at best are disfiguring or, at worst, can cause such severe structural imbalance that it may uproot them. Which is why if you happen to have evergreens growing in your garden, help them out where practicable by relieving boughs bending under heavy snowfall with a good shake or a broom.

Shrubs are a thrifty lot, and like cats, they scope their surroundings, always ready with a contingency plan. However, with so many species working at very different levels of winter dormancy, shrubs need a superior game plan guaranteed to give them a head start in the annual spring bunfight. Well aware that once the weather starts to warm up, they face stiff competition from neighbouring plants, all jostling for position, any tree or shrub bearing dangling catkins on bare branches deliberately flower early, knowing that they have to make the most of available sunlight before surrounding plants put in an appearance and rob them of the sun's warmth.

Many beautiful deciduous shrubs, including witch hazel, silk tassel bushes and viburnums, smartly prioritise reproduction. And like squirrels, they wisely start planning early by accumulating carbohydrates and storing food through the winter months. Instinctively knowing that both flower and leaf production require sunlight at a time when sunshine supplies are low, they err on the side of caution and ensure they've laid by adequate supplies of carbohydrates to fuel their growth. Then they take a no-nonsense approach to procreation by blooming early on bare stems. Once the flowers have done their thing, they can safely divert their energy back to leaf production.

Naturally, marketing has a huge influence on the plants we choose to enhance winter's tedium by offering enchanting shrubs that are guaranteed to bloom in the iron depths of winter. Unashamedly oriental in origin, camellias are rightly touted as free flowerers, along with winter-flowering pansies that can cope with mild winters but will droop or stubbornly refuse to bloom in extreme cold. While undeniably beautiful iced with snow, acid-loving camellias haven't quite worked out that warming morning sun will torch their flowers beyond repair and hope gardeners will have the good sense to plant them in more hospitable aspects. North and west-facing spots are best, thus avoiding frozen blooms turning to mush. Easily living up to 200 years and first arriving in Europe in the 1700s, shipped by the East India Company, camellias number over 200 species, including small trees alongside numerous shrubs. Overshadowed by more compelling varieties, the most famous member of the family is rarely accorded ornamental pride, yet certain camellias, populating coastal areas and mountain groves from China to Japan, account for over 5 billion cups of tea drunk globally every day.

Some plants are born chancers, relying on opportunity in their fight for survival. And you'll be astonished at the speed at which our most-hated garden plants can grow. Weeds.

WEEDS:
Chancers and Opportunists

Spend all morning weeding your patch and the moment your back is turned, weeds sprout again! Damned things are everywhere. Champion subsisters, I daresay if they had to pit their survival skills against herbaceous plants, they would win hands down. Of course, any plant could be considered a weed if growing in the wrong place, but why are weeds far better at expanding their numbers than cultivated plants?

The success of lovable literary characters like Jeeves and Wooster pivoted on P.G. Wodehouse's assumption that the British gentry weren't blessed with a great deal of natural intelligence, so minions like the inimitable butler Jeeves did the critical thinking for them. Nature has its own curious version of levelling the playing field and it's absolutely mind-boggling once you start sifting through a weed's capabilities.

Weeds are journeymen possessed of cast-iron fickleness that enables them to settle swiftly in new environments, get their roots down sharpish, reproduce and spread at lightning speed. On poor, compacted ground or newly turned soil, sand or spoil, a dormant weed seed lying in wait leaps at the least opportunity and even chopped roots invariably form new plants. Weeds come in two distinct flavours – annuals and perennials. Let's meet the annuals first.

Completely unfazed about comfort ratings, annual weeds are fast learners, adapting to whatever lodgings are available. Believing the welcome mat is laid out all year round, they don't know the meaning of two's company and three's a crowd and certainly are no respecters of boundaries. Growing so fast, they don't bother diverting much effort in developing intricate underground storage organs or complicated roots; they are on a mission – speed dating and baby farming, and if there are no decent suitors out there ... hell, they'll just DIY!

Although pollinators such as bees and butterflies might visit purslane's golden flowers, being self-fertile means most of its blooms can automatically produce seed without outside assistance. And we're not talking about a few scattered handfuls of seed either. A single purslane can produce over 52,000 seeds in its short lifetime, and approximately six weeks is all that is required to hatch, match and dispatch.

Originally from North Africa and the Middle East, purslane populates cracks in paving, sidewalks, fields and scrubland, and is deliriously happy residing near gas-spouting motorways, where the habitat is far from nurturing. Being rich in antioxidants, purslane has even sneaked into our kitchen herbary (I hope you enjoy it because it's the nearest to victory you're likely to get). If seeds don't get lucky this year because the weather is too hot, too dry, too cold or too wet, well, heck! Ripening swiftly, pods split open, ensuring the next generation are unleashed on the world in short time. And uprooting a purslane plant when it's studded with seed capsules won't deter the pest one bit. The roots may die, but their pods live on, cracking open at the first opportunity. As seed can remain viable for years, one way or the other, like other annual weeds, purslane will survive.

Aside from a rapid growth spurt, other ingenious coping strategies include built-in drought resistance via fleshy leaves that act as water storage, and so purslane isn't too fussed whether it rains or not. In periods of drought, it simply slows its growth rate, rations water until the next rainfall arrives and then it's off the starting blocks once again.

Bindweed, chickweed, crab and goosegrass will rise up and challenge anything in their path. Shrubs, tumbled-down walls and fencing, tree stumps and rocks pose no hindrance to their ambitions because they produce a staggering quantity of seed compared to gentler herbaceous plants.

Not technically weeds but wildflowers, common poppies might be more endearing but they, too, fling out over 1,000 seeds per capsule, so you can see why annual weeds play the percentages. Spread enough around, and there's a good chance some of it will make it to adulthood. Ducks have a similar laissez-faire approach in raising their offspring. Ducks, you say? Bear with me.

Typically, mallards have clutches of around a dozen eggs or so, compared to swans, which lay perhaps half that number. While male ducks are content to guard their females during courtship, once they've had their wicked way, they disappear pretty sharpish when it comes to shouldering the responsibilities of parenthood. Having been a duck-keeper, I'm sorry to inform you that mother ducks are also often found wanting on the parenting front. Laying numerous eggs, the moment they hatch, she plunges her ducklings into the nearest water bath without so much as a backward glance. All that shrill peep-peeping is the sound of youngsters trying to catch up with mum, who is already shamelessly fluttering her eyelashes at some nearby bobbing drake. With no intention of sharing watery acreage, moorhens, who are notoriously territorial, sneak up on any vulnerable ducklings that have drifted dangerously from mum's wake and ruthlessly dunk them under water until they drown. Swans, on the other hand, protect their eggs with savage ferocity and both parents stick around to raise their young, so they can afford to limit the number of eggs laid.

Weeds such as hairy and wavy bittercress provide food for wildlife and thus, we are instructed to treat them gently. But a gardener has only limited patience. The annual version is capable of reaching maturity in as little as twelve weeks, and the latter can't make up its mind whether it a biennial or a yearly nuisance, but at the rate both grow, it matters little. Pretty much growing all year round, only a severe cold spell will break its stride. But not for long. A single plant can foster up to 1,000 seeds, which they sling in all directions for several feet. Hoe them all you like, look away for a second and another will pop up in its place as surely as duck eggs are duck eggs.

Admittedly, their tiny white flowers sustain beneficial insects – they are even pretty if you squint hard – but sometimes difficult decisions have to be taken. There isn't room for rivals in a garden, so it's down to you to choose which plants get shown the door. Luckily, bittercress are fairly shallow rooters and a firm tug condemns them to the compost heap; just make sure you don't let them flower first, or they'll contaminate the whole pile.

In recent years, it hasn't escaped my notice that even the Royal Horticultural Society have abandoned calling weeds ... well ... weeds. That old gardening wisdom that a weed is simply a flower in the wrong place sounds all very civilised on paper, but in real life, I think we *all* know a weed when we see one.

Setting up permanent homes and blooming each year, perennial weeds including bindweed, chicory, Japanese knotweed, perennial nettles and plantain are in it for the long haul, bringing up their young with patient persistence. Cut a perennial nettle down to ground level and underground its sinewy mustard roots continually spread and conquer new territory. Snow, frost and ice may decimate a nettle's upper parts but its roots snooze unaffected underground, releasing water from their cells to prevent them freezing before shamelessly rampaging again in more clement weather.

Lawn weeds have no such compunction. Frustrating even the sharpest blades, mow all you like, they are structurally immune, having adopted a low, spreading habit where they are less likely to sustain injury. Perennial dandelions have deep, branching roots that dive down to depths of 2ft or more, challenging the most determined hoe or weed torch. Nine times out of ten, you may think you've managed to pull up the entire plant, but if a small section snaps off in the struggle, it grows into adulthood. Dandelions' fluffy clocks fair relish the threat of cold steel at their throats. Chop off their heads and their feathered youngsters are suddenly airborne, free to drift and inhabit any place of their choosing.

Japanese knotweed. Oh dear. What a shame it's so invasive because it's a pretty decent-looking plant. It's a formidable intruder, known for supposedly being able to break through concrete and tarmac, but as with many myths, it's all smoke and mirrors. Knotweed doesn't have tissues tough enough to break solid concrete, but it will reliably exploit any weaknesses in hard landscaping, crumbling grout between paving slabs, cracks in walls or mortar in brick joints, and spreading its root run in feet rather than inches. It must like even numbers since it can grow 10cm a day with ease, which explains why it tops just about every public enemy list globally.

If you've ever battled with pernicious mare's tail, you'll know you're up against the Voldemort among weeds. I swear the only thing that will shift it is a hefty dose of napalm. When I first tripped across this piratical usurper in a corner of my garden, I hadn't a clue what it was, much less its megalomania. Since it's a very good-looking individual resembling a low-growing thicket of mini-conifers in its youth, I left it alone. How was I to know that a few months later that small emerald patch would aspire to acreage? If you have no Agent Orange handy, follow my advice – digging is absolutely pointless. Any root fragments breaking off in the struggle readily form mini-plants, so avoid pulling up stems for the same reason – snapped stems underground make new plants.

It comes as no surprise that mare's tail predates the dinosaurs – a genuine living fossil – so to be rid of it, you need to think like it. Use what we've learned about conditions in the dinosaur days and use it against them! We know that soil was boggy, and oxygen and nutrient levels were low, all signs of acid soil. (Okay, dry soil can be acidic too, but the same methodology applies.) You'll need to ramp up the pH (or soil fertility to the uninformed) and drastically improve drainage by installing pipes or filling dents and holes with earth and stones so water can more easily drain away. Changing the soil dynamic sends a signal that climate change is happening in that part of the garden, and rather like their old mates, the dinosaurs, mare's tail has no option but to die out. To finish, add a handful of garden lime, which is ideal for balancing soil fertility – follow the manufacturer's instructions, water it in and wait. You're on the home run! But take your eye off the ball for a second and you'll be back at square one.

On the next leg of our journey, we're meeting a fascinating tribe with survival instincts that make weeds look like a bunch of amateurs.

PART FOUR

INIMITABLE LIFESTYLES

Nature is the difference between the possible and impossible.

– The author

Plants aren't all just leafy, green sunbathers. They share the same basic job – turning sunlight into energy through photosynthesis – but they don't all live the same way, and their lifestyles vary as much as their shapes and colours. Some are adapted to fertile habitats, some prefer the damp shade of a rainforest, others grow in arid deserts. There are plants that climb, plants that float, and even those that eat rodents. Each species has its own inimitable lifestyle, shaped by the environment it calls home.

CARNIVOROUS PLANTS: The Most Marvellous Plants in the World

Described by Darwin as the 'most marvellous plant in the world', the Venus flytrap resembles an alien species – all neon-green fringes with curious needled spines. Twin clamshells gape with open invitation, tempting insects to drink the honeyed nectar lining its inner walls. But lurking unseen in the plant's mouth are fine hairs that are ultra-sensitive to visitor movements. Small gnats may trickle through its interlocking spines unharmed and drink their fill, but the moment a Venus flytrap senses a larger intruder – wham! In under a second, the leaves slam shut, trapping the unsuspecting visitor in an unrelenting prison. The more an insect struggles, the tighter its embrace and suddenly that magical nectar is transformed into a fatal acid full of meat-decaying enzymes. Over several days, the flytrap steadily releases a lethal bile until it finally absorbs its protein-rich meal. The observation at the time between Darwin, his friend Dr Gray and a fellow enthusiast, William Canby from Wilmington, Delaware, records the moment of revelation:

> Darwin has hit it. I wonder you or I never thought of it ... Think what a waste if the leaf had to go through all the process of secretion, etc., taking so much time, all for a little gnat. It would not pay. Yet it would have to do it except for this arrangement to let the little flies escape. But when a bigger one is caught he is sure for a good dinner. That is real Darwin! I just wonder you and I never thought of it. But he did.

Other carnivorous plants have less dramatic, but no less effective dietary habits loaded with cunning. Some display vivid-coloured organs to trick prey into thinking they are actually blooms; others use camouflage to blend in with their surroundings to dupe their victims. Some even flower at the tips of tall stems, wisely keeping their blooms as far away as possible from their deadly digestive organs so they are able to distinguish between beneficial pollinators and their next meal.

It's easy to see why carnivorous plants captured Darwin's attention when you witness the lethal tactics of the world's largest carnivorous species – the pitcher plant, quite capable of digesting small mice, frogs and reptiles as well as insects. In the heat of summer, small thirsty creatures are attracted to liquid in the pitcher's bowl, which can easily hold 3 litres of digestive fluid. Mistaking it for water, a mouse clambers onto its slippery rim and as the unfortunate rodent bends forward to take a sip, it topples forward into a deep urn with no means of escape. Suffice to say, pitcher plants make malodorous neighbours as, over the following weeks, its victims are slowly digested. Bats, on the other hand, are unperturbed by the threat, quite commonly using its lids as shelters.

Sundews, while summoning up visions of plants drenched with sunny appeal, are equally deadly in their method of survival. Another Darwinian favourite, commonly found growing in acidic habitats where essential nutrients are hard to come by, sundew thrives at the edges of boggy pools, soggy heathlands and peaty moors. Small lime-green cushions dotted with multiple wine-coloured hairs are beaded with enticing dew. When unsuspecting insects alight on its gluey beads, mistaking the gleaming drops for nectar, the movement triggers curling tentacles that clamp prey in a fatal clasp.

Interestingly, many carnivorous plants aren't as interested in eating their prey as attracting their attention. The cobra plant, so named for its snakehead likeness, is incapable of producing any digestive enzymes at all, inviting birds, bats, mice and lizards to defecate in its chambers, their droppings providing the vital nutrition needed to power the plant. To maximise the chances of a full larder, that distinctive hooded, patterned head is also skilfully designed to trick insects into entering its curiously bowed throat. Once trapped, it's literally the slippery slope as victims hurtle to their deaths, providing the life-giving enzymes that allow the cobra plant to digest its food.

A protein-rich diet is all very well, but such heavy sustenance demands a respite, so carnivorous plants take a break for three or four months over winter to recharge their batteries. As the days grow shorter, coinciding with a depleted food supply when insects are less bountiful, even meat-eaters are forced to slow down, hibernate and wait for warmer spring weather.

In contrast to the cobra plant, common bladderwort couldn't look less threatening if it tried. Lurking below the surface of ponds, lakes and shallow ditches, and with no real roots to speak of, bladderwort forms a tangled green mesh studded with small, rounded sacs or bladders. Throwing up slender stems of golden flowers above the water line in summer distracts us from more sinister happenings below. The bladders are ingenious mechanisms fitted with watertight, spring-loaded trap doors bristling with ultra-fine filaments that filter pond life, including mosquito larvae, tadpoles and water fleas towards it. Once the swimmers come in contact with the hairs at the entrance, the bladders collapse like balloons, and the incoming rush of water carries the bathers into a watery dungeon, slamming the door behind them. And all this happens in less than a millisecond! Deprived of oxygen, within hours captive paddlers create a nutritious appetiser for the plant, which is already thinking about the next course. It immediately resets, pumping out any excess water, and in under twenty minutes is ready for its next victim.

Cycads have no such hurry. Benign and patient, they enjoy an altogether slower pace of life combined with an impressive heritage that has allowed them to endure for centuries.

CYCADS:
Ancient Warriors

Better known as Japanese sago, cycads come from an ancient line of cone-bearing plants, and stood, if not shoulder to shoulder, then toe to toe with the dinosaurs, roughly 12 million years ago. As the Earth's climate shifted and clear seasons emerged, across Australia, Africa, Asia and Central America, these resilient evergreens pushed into new, challenging habitats in a constant struggle to survive. Numbered among the world's most endangered species, these ancients adapted and flourished and still stand defiant in the face of millennia.

If you happen to live in warm, temperate climate, I daresay Japanese sagos are probably just another average garden plant to you, but to a British gardener denied the luxury of growing them outdoors because of our inhospitable climate, they are exotic, mysterious beings. Striking good looks teamed with undeniably beautiful architecture – deep green fronds fanning out in perfect symmetry – make them a landscaper's dream. Incredibly slow growing and rarely suffering from sickness, my first encounter with their mating habits threw me in at the deep end – an astounding ritual conducted at such speed and precision that had I not seen it with my own eyes, I'd never have believed it.

Having grown half a dozen of these strapping fellows on my terrace for many years, to be honest, I was so seduced by their contemporary swagger that it never occurred to me ponder what sex they might be. A savage late-summer storm thrashed the pots, leaving the unfortunate Cycads hanging half in, half out of their containers – time for rehoming. Easier said than done, with each weighing more than a dead donkey. Between my partner and I, we improvised a stretcher, a half-inflated lilo – the only thing to hand – and tying up their spiky fronds, we dragged them to the shady rear garden.

Then we set about the back-breaking task of digging holes that might have easily accommodated a dead relative or two. Reasoning they were less likely to lose water in the shade, and thus giving them a better chance of recovery, and with fresh soil accompanied by a good soaking, I crossed my fingers and hoped they just might survive the upheaval. To my horror, the following morning, the fronds had turned a dirty mustard yellow and hung like dead men on the gallows. We'd killed them! Then, I noticed something strange ... Stay with me, because this is where it gets interesting.

Overnight, one plant had sent up a pineapple-sized cone at its centre. The others, which the previous day had rounded, fist-sized middles, suddenly ballooned to twice their size. Clearly, it was a response to upheaval, they were trying to save themselves.

Stuck on an isolated Mediterranean hillside with no recourse to the internet to explain this sudden growth spurt and forced to make my own deductions, it was easy enough to guess they were probably wind- or insect-pollinated. Being late in the year with precious few insects around, I was suddenly cast as an unlikely matchmaker.

Over the next three days, the same thing happened. Each morning, those centres doubled in size. By the end of the week, the pineapple stood easily 2ft tall while the others had developed feathery middles the size of footballs. Only then did I realise I must have a mix of male and female cycads – one with an enormous spike that I guessed was male, and the others with curious furry scales. An emergency dash to my neighbour revealed I had roughly a week to hand-pollinate them or they'd be in danger of turning up their toes permanently. So, I set about detaching the surprisingly heavy male spike, which released an extraordinary cloud of pollen onto a sheet of newspaper. Carefully, I leaned over each female football, whose delicate scales had loosened, presumably to allow insects entry, and liberally dosed each with pollen. Their fate was in my hands.

Was it too little, too late? No longer interested in my clumsy romancing, within a week the females closed their scales, and smart new fronds replaced the sickly ones. Thankfully, my sagos surpassed their former finery, but the rate of recovery was nothing short of astonishing. Having been forced to become part-matchmaker, pollinator and ally, this botanical episode made me realise just how exacting nature can be. Oh, and of course, that timing is everything. And since cycads are commonly mistaken for palms, let's take a look at why that might be.

PALMS: Neither Fish Nor Fowl

According to definition, trees are woody plants with a tough outer covering of bark, and since palms lack this woody overcoat, according to botanical definition they fall into a bit of a grey area. Long lived, surprisingly hardy and flourishing in diverse climates from Northern Hemisphere suburbs to sweltering Floridian coasts, to the casual observer, palms possess tree-like characteristics by way of a single trunk topped with a canopy of leaves – but they have grass-like features, too. So, why aren't they trees?

Plants are divided into two broad categories, monocots and dicots, based on their physical features and growth patterns. Dicots have a trademark woody main stem, and their leaves grow in multiple layers as seen in oak, beech and conifers. Whereas monocots (palms fall into this group) begin as sprouting seeds with a single leaf that looks more like a blade of grass and shoots up quickly or more slowly depending on its habitat.

Palms never grow bark. Instead, their trunks are made of hardened, fibrous layers formed from previously spoiled fronds that are sturdy enough to support the weight of new foliage. And unlike woody trees, they lack the ability to produce the magical substance cambium, an essential multipurpose tissue devoted to plant growth and regeneration. Trees manufacture cambium in plenty so can self-heal wounds or damaged and broken branches, but palms have no recourse to this vital built-in first aid and once wounded their wounds stay with them for life.

Fortunately, palms are blessed with an important versatility to weather any storm. Saw through any woody tree and you'll find those concentric medullary rays – you won't find these in palms. Instead, they have circular vessels spread evenly throughout the trunk and it's these scattered vascular bundles that make them both resilient and pliable in violent storms, often seeming to touch their toes in hurricanes without incurring serious injury. An enviable agility that usually results in breakage, limb loss or death in more rigid, barked trees.

The hurricane palm (also called the princess palm) is an expert storm dodger, dropping its leaves the moment it senses ferocious winds, and since the winds can't catch its fronds, it's rather like harassing a beanpole. The princess palm waits patiently until the storm subsides, surviving another herculean assault naked, but unbowed.

However, all that coarse matting made up from spent fronds means there are some elements palms fail to repel. In recent years, tens of thousands of hitherto healthy palms have met premature death owing to the stealthy but deadly red palm beetle. Attracted by the smell of sap when a palm is cut or pruned, females lay eggs in their craggy crevices and the voracious larvae gorge on the inner soft tissue, eating it from the inside out until the palm gives up the fight and either rots or is felled.

Like much of our threatened flora and fauna, some palms are having a difficult time of things and sadly their numbers are dwindling. Where they have tried to adapt to stressful conditions and still haven't managed to thrive, happily, human intervention has seen rare species flourishing again with unexpected benefits. The Californian or desert fan palm, found in isolated locations between the Mojave Desert and Colorado, needs very specific living conditions. At approximately 40ft tall with an impressive girth of around 15ft, its sizeable hanging skirt comprising dead fronds might look unsightly but gives shelter to countless creatures – bats, birds and reptiles.

As it is largely a desert dweller, water is understandably hard to come by. Unable to up sticks and relocate to a more favourable environment, it has unexpectedly mitigated the odds through human demand for its huge, lush fronds in domestic landscaping. Should you live in Hawaii or Florida and be growing a desert fan palm in your backyard – congratulations! You are unwittingly contributing to its survival. Quite by accident, it has managed to expand its numbers by naturalising in areas where gardens are cultivated and agriculture prevails. And where's there's agricultural activity, there is water, so they have happily landed in areas where warmth and abundant water provide a real chance at life. While many of its palm brethren have been decimated by the red palm beetle, the desert fan has demonstrated resilience where other warriors have fallen and, unusually, it is able to repel deadly weevils by secreting a gummy substance that kills off the larvae.

If you are seeking armour-plated defences, you might consider the elaborate fortifications displayed by Mexico's chocho palm. As interested in reproduction as survival, sharp spines clad both the bark and foliage to deter even the most determined marauders. Even its flower pods are rigged with vicious spikes, and its seeds might as well be wearing Kevlar vests. It is preyed upon by rats and, enchantingly, a harmless-looking predator: the spiny pocket mouse, with a tail longer than its body length. Those fearsome spines turn out to be as threatening as human body hair, but this cute little Disney mouse must have some seriously nifty footwork, for it is the only creature that comes close to nibbling the chocho's impenetrable seeds.

But why some palms make it so hard for potential pollinators to get at their seeds is perplexing. Enter the fickle traveller's palm. Resembling a giant Incan headdress, a mature palm is a sight to behold. Stretching upward at 12m or more, you could hardly be blamed for thinking it should qualify as a tree. The name is misleading, and I suspect the misnomer has imposed an identity crisis because it isn't a palm at all, just another flowering plant related to bananas, cannas, ginger and the exotically plumed bird of paradise. Plain, sturdy, straw-coloured flowers lack any real allure, a disadvantage more than made up for by generous measures of nectar held in inaccessible deep chambers.

And here we hit another puzzle. Impervious to sunshine's enticement, its flowers remain obstinately closed, so it's mighty odd a plant would hinder its chances of reproduction through sheer bloody-mindedness. Madagascan lemurs to the rescue! More precisely, black and white ruffed lemurs, the palm's preferred pollinators. Leaping with enviable agility and foraging in treetops for fruit, seeds and leaves, and living on average around twenty years, ruffed lemurs spend their days in tight-knit family groups and can charm open the traveller palm's blooms with consummate ease. Endearingly amber-eyed and inquisitive, lemurs have these haughty virgins all figured out, and being critically endangered themselves, never was a symbiotic relationship more vital. Fortunately, lemur snouts and the palm's resistant blooms are ideally matched. Lemurs pry the flowers open to get at the sweet nectar hidden inside, inadvertently dropping pollen onto the stigma and leaving the plant resoundingly fertilised before bounding off in search of other favourable food outlets. And because lemurs conveniently sense green- and blue-shaded spectrums, the palm's electric blue seeds cement this intimate relationship further as they help scatter seed throughout the neighbourhood.

The African raphia palm, responsible for raffia twine and all kinds of woven baskets, deserves a thought as you tie in your roses. Since it lives in areas increasingly threatened by habitat loss, it has figured out the only way to get ahead in life is by growing taller than everyone else. Given that its short stem is buried underground, it's something of a miracle that it ever manages to tower over its neighbours at all. Featuring massive fronds easily 80ft tall and 10ft wide, with an estimated weight of 100kg, it would need half a dozen people to carry a single leaf. Its appetite for sunshine is fully justified.

After growing for years, it undergoes a sudden growth spurt, flowering just once. Raphia's male and female impressive reddish blossoms fade to huge, burnished-copper columns of seed and, having fulfilled its destiny, it keels over and dies.

The accolade of the world's tallest palm goes to the Quindío wax palm with heights up to 200m and, you'll be pleased to know, the world's smallest palm can be grown in a home terrarium or a pot on the patio. Aptly named *Dypsis minuta*, it attains a modest height of around 50cm, and depending on conditions and with the right care, it might well outlive you.

Being benevolent creatures, I don't suppose palms think very much about n'er do wells, and despite their superior strength, plants often find themselves up against some pretty undesirable scroungers.

PARASITES: Freeloaders and Scroungers

Armed with nothing more than a whip and jaunty kerchief, Indiana Jones famously thrashed his way through dense jungles and arid deserts in search of long-lost treasure. Surviving colossal boulders barrelling towards him, terrifying waterfalls and risking being buried alive in underground crypts, the baddies always got their just deserts, and Indie always got the girl. Huzzah!

Extreme locations such as those depicted in movies are home to a great number of plants that have gained celebrity through shameless freeloading. Unable to hack it alone, they steal from others and largely exist in a world of human indifference. Famously elusive and curiously beautiful, some parasitic plants are wholly dependent on other species to survive, while others are less harmful to their hosts, more akin to scrounging relatives.

A plant that takes some – but not *all* – of its nourishment from a host, for example, mistletoe, is known as hemiparasitic. Of the thousands of plants on the planet, only about 4,000 plants are fully parasitic (obligate parasites) and it's thought that roughly 76 per cent have never been cultivated. Their rarity makes them tricky customers to track down, ergo many of their number are in decline and, unsurprisingly, you have to find one to grow one. To be honest, generally, they aren't greatly blessed in the looks department, which hasn't exactly elevated their popularity. Still, they're a fascinating bunch and deserving of closer inspection.

The distinguishing feature possessed by all parasitic plants is a modified root, a special organ called a haustorium, which punctures the surface of a host plant and plumbs into its vascular system. Take Maltese fungus – which isn't a fungus at all, it's a misnomer – like most fully parasitic plants, it's a peculiar-looking individual found in hostile, salt-thrashed coastal habitats. Also called desert thumb, it is a holoparasite, totally incapable of assimilating carbon and nitrogen, unable to photosynthesise or produce chlorophyll and is therefore forced to steal nutrients from the roots of surrounding plants for survival. The only time it makes an above-ground appearance is during the rainy season. Until then, it stays busy, building up its energy to power its impromptu appearances. Roughly the size of my hand and jutting gherkin-like in the sand, it produces cones studded with hundreds of tiny black-red flowers resembling compressed potpourri but not nearly as fragrant. A rotten cabbagey waft tells you its preferred pollinators – flies. Which makes our sandy nomad sound rather unappetising, doesn't it? But never judge a book by its cover! Maltese fungus has gained sex-symbol status, being greatly valued in treating male impotence and sexual dysfunction.

Toothwort, a woodland dweller, has a similar MO. It's an unlikely looking con artist that's also bereft of the required skills for sole survival, stealing food and water from plants such as alder or willow. Gorging underground for most of the year at its host's expense, you might think toothwort has shot itself in the foot when it comes to getting fertilised. But this scheming glamour puss knows its mark. Fleetingly flowering for around four weeks around mid-spring, she's all dressed up with somewhere to go. Laying down a carpet of stubby, hooded violet flowers, she wafts her signature ammonia scent from deep nectar chambers. The unpleasant whiff, while off-putting to foraging birds, allows her to save her favours for her preferred pollinators, bees or any insects equipped with extra-long tongues to service her needs. Then – just as suddenly – she disappears underground again! But it's opportunity enough.

Woodlands abound with mysterious plants. Venture into any beech or conifer wood and you'll encounter the ghostly hoods of Indian pipes. You could be forgiven for mistaking their pale clusters for some kind of etiolated fungi. Completely white or translucent pink from head to toe, showing only malevolent, black eyes, these wild perennials can't even be bothered to do the mugging themselves. Instead, they stalk friendly fungi already in meaningful relationships with existing trees, creating a curious ménage à trois. Unable to make food, since they lack chlorophyll, Indian pipes steal sugar from trees via ensconced fungi, offering nothing in exchange.

Mummy may have been caught kissing Santa under the mistletoe, but our favourite festive greenery has an ill-gotten reputation that all begins with a kiss. Less deadly than strangler figs but mightily inconvenient for a tree at least, well-intentioned sticky seeds are innocently scattered through the landscape via bird droppings and lodge in tree branches where they germinate. As a mistletoe seedling grows, it pierces the tree bark, obstinately latching into softer tissues with that covert but ingenious engineering adjustment haustorium and sneakily siphons off unlimited food and water. It's a one-sided relationship (we know all about those, don't we?), forcing the tree to share its water supply and depriving it of vital nutrition. Since mistletoe is a long-lived perennial, the tree has no way of shaking off its freeloading lodger and is stuck with its mean-spirted co-habitant forever and a day. In fact, some plants have got so good at leeching that they have abandoned any notion of fending for themselves, relying on that invidious bridge that inextricably links them to their victims.

While we're in festive mood, let's meet a notorious ligger from Down Under. Scrounging in similar fashion to mistletoe is the Australian Christmas tree. It doesn't intentionally set out to become one of life's cadgers but, if you happen to be living next door to one, they are not to be trusted. A rangy Western Australian native, blessed with distracting clouds of vivid orange blooms, looks are deceiving – it'll take you for everything you've got. Quite capable of photosynthesising and paying its own way, unfortunately an ingrained kleptomania drives it to steal water from surrounding plants. The Aussie Christmas tree ruthlessly taps into a host's roots with its razor-sharp cutter, severing water, food and mineral supplies – and often fibre-optic cables and power lines while it's about it.

Some of you know that once you get an invasion of dodders, they're a hard tribe to shake off. Dodders are obligate parasites, which are unable to thrive or reproduce and sustain themselves entirely on the host plant. To the uninitiated, it's hard to tell bindweed and dodders apart, both coiling raiders that are so widespread that they have been given any number of nicknames, none of them complimentary. Devil's gut, strangle weed and wizard's net – take your pick. But if you thought bindweed was a sod to get rid of, dodder takes the biscuit.

Using the same cutting technique as toothwort and attaching to their host via wiry tendrils, what singles dodders out from the common herd is their ability to sniff out potential victims. Highly susceptible to infant death syndrome, a young dodder has a short window of opportunity – usually about a week to find a host before kicking the bucket. Seedlings smell the chemicals given off by unsuspecting targets, and once infant dodders pick up the scent, they instantly change direction, growing towards their intended victims – plug in their cutters and they are set for life.

I'm sure you'll have noticed perfumed cut flowers such as lilies smell more strongly as they decline. Strangely, it's not unusual for flower scent to increase as blooms fade. Detached from the mother plant, flowers know it's their last chance to procreate and with those last desperate gasps of perfume, they are still hopeful of attracting pollinators to sustain their legacy.

Let's briefly revisit the corpse flower. A massive, red-speckled affair, lacking leaves, roots or stem, rendering it unable to photosynthesise and forced to take up residence inside nearby tetrastigma vines. Stealing nutrients and water as it works its way towards its magnum opus, it grows out of the vine's bark in the ebbing days of its life and diverts all its effort into producing one massive flower – a great lobed affair, studded with curious pale markings around a gaping central crater. Weighing in at around 10kg, the single bloom releases the scent of rotting meat for just one week in the desperate hope of attracting pollinating flies before dying.

Quite why some plants take such lengthy contemplation in securing their successors is a bit of a mystery. However, others are less scheming in their approach and it's all too easy to tar parasitic plants with the same brush.

Yellow rattle, a pretty, yellow-flowering meadow plant, is easy to identify because it's usually foaming with cuckoo spit. With a hemiparasitic nature, it syphons off some, but not all of its host's nutrients. And it's pretty sharp at turning the tables on surrounding aggressive grasses by exerting traction at its competitors' roots with that reliable cutter, depriving them of food at source and thereby creating opportunities for some other, gentler species to thrive. Once its flowers fade, they are replaced by charming pouches of rattling seed, spread by wind, walkers and other hairy creatures.

Just as fascinating are trees – we could devote several volumes to their faithful service. So abundant in number and manifesting in all shapes and sizes, it's all too easy to take them for granted or mistake them for shrubs.

TREES VS SHRUBS: Why is a Shrub Not a Tree?

Let's skip back to basics for a moment. It's entirely forgivable to confuse shrubs with trees because some shrubs can get very large indeed. Take Portuguese laurel, a handsome evergreen with appealing wine-coloured stems, a white froth of flowers bursting from glossy, deep-green foliage and, more often than not, offered as hedging. However, with the potential of reaching a mature height of 40ft, it's not always evident if it is a tree or a shrub (although, I daresay laurel is happy to be mistaken for a person of substance). To complicate things further, there are a number of small trees that don't grow nearly as tall as some larger shrubs, so it can be confusing to the novice where a shrub ends and a tree begins.

Generally, it's safe to say that trees are woody individuals with a main central trunk and a recognisable canopy. Shrubs, on the other hand, are simply woody perennials with multiple stems growing nearer the soil line and tend to be larger – although not always – than herbaceous perennials. Much as they would like to, not all shrubs or trees reach their advertised heights since available water, sunlight, soil and where they are planted all have an effect on their eventual stature. Shrubs are relatively quick growing and can reach maturity within five years, whereas trees can take anywhere between ten and fifty years to reach full majesty. However, often ultimate size becomes irrelevant if they are particularly fast growing or have been planted in a very small plot.

I hadn't intended to devote many words to shrubs. However, they deserve a mention in their own right since you couldn't hope to meet a more dependable bunch. For the most part, shrubs are tolerant and good tempered, and as you become more experienced in your gardening practice, you'll become more attached to them for that very reason. As a horticultural student, I was hopelessly seduced by the dizzying line-up of glamorous perennials – most of them new to me at the time – and it wasn't until some years later that I began to appreciate the kaleidoscope of beautiful shrubs on offer. That's not to say they don't count a few prima donnas in their number, but nine times out of ten, a shrub lets you know where you stand, which in my experience isn't always true of human nature.

Indulge me if you will as we trespass into the hallowed world of rhododendrons. A Marmite crowd if ever I met one, attracting devotees and naysayers in equal measure. Usually bundled up with their smaller relatives, azaleas, folk rave about rhododendrons' explosive spring colour and you'll be pleased to learn there's a growing number of summer bloomers hitting the market now. In case you're wondering, rhododendrons and azaleas are one of the largest plant families in the world with well over 1,000 species. You can always tell the difference between the two: azalea have smaller, pointier leaves than their loftier evergreen relatives and are smooth to the touch. But such is the fevered reverence in which 'rhodies' are held, gardeners will insist on growing them no matter what.

The thing is, when it comes to soil, rhododendrons are a pernickety lot. Being acid lovers (or lime haters, depending on your point of view), they simply won't put up with common clay, sand and – horror – they've even been known to cock a snook at coveted loam. Emanating from China, Japan and the Himalayas, what rhodies like best is well-drained, peaty soil and they'll accept nothing less. If your garden is surrounded by beech, oak or conifers, there's a good chance you can grow them successfully, but if you're not blessed with naturally occurring ericaceous soil (soil with high acidity), I'd give them a miss. Unless, of course, you're prepared to retrain as a barista and spend arduous hours blending additives into your soil, and even then you'd need a ready supply of chopped bracken, pine needles, sulphur or chelated iron to accommodate their exacting needs. If you're a bit of a chancer, you might get gung-ho about the whole caper and shove them in whatever garden soil you happen to have and for a few months you might well think that you've defied tradition. Just wait. Their leaves inevitably turn a sickly yellow and those opulent flowers become mean and accusatory, sneering with disdain every time you walk by. And whilst we're advised to give them a *particular blend* in which to root their capricious tootsies, I'd hazard an educated guess that drainage is just as important a factor – any acidic, fertile soil with the benefit of added gravel or rough sand should see good results.

Frankly, as a kid, I was less impressed by their extravagant blooms than their capacity for adventuring. Our rhodie thicket was a veritable plantation, ideally suited for camp building. Every clique in the school built their own bespoke camp under their dense, leathery canopies that, even in downpours, afforded the luxury of letting you stay relatively dry. Distracted by their opulent blooms, much overlooked by the gardening fraternity are the recreational possibilities provided by their multi-stemmed architecture. Easily climbable, rhododendrons are ideal for ambushes, perching, swinging, hanging, lounging and hiding. The rhodie camp became our personal Kirrin Island where we re-enacted *Famous Five* adventures, composed pop songs, play-acted, carved penny whistles from brittle branches, hid from unjust detentions and regularly raided neighbouring camps for interesting booty. But our prize pillage was an old wooden barrel with rusty banding, perfect for tormenting snooty day girls. This item was reasonably shared between rival camps and if it's of any interest, simply add a handful of sand to the barrel, abduct a suitable transgressor, stuff them in the barrel and then, with the help of your co-conspirators, roll them down a nearby slope until they scream for mercy. They normally emerged humbled, mostly huffy, but always gratifyingly itchy and red-eyed, but it sent out a clear message that boarders were not to be messed with. Besides, it was rare any teacher would come searching Rhodie Wood for miscreants and, even to this day, I've yet to be convinced that their sumptuous flowers hold a light to their outward-bound potential.

Like many plants when threatened with extreme weather, these broad-leafed evergreens think nothing of curling their leaves into drooping cigars. Thick, insulated leaves can cope easily with temperatures as low as 20°F/ -7°C, so why shrink when they are adequately equipped to protect themselves from extreme cold? We know that water in leaves is prone to freezing, often resulting in irreparable cell damage, but as it turns out, it is more about the rapid thaw that follows. Frozen, curled foliage ensures the plant doesn't receive as much sunlight if the sun makes an unexpected appearance, thus slowing the thawing process and minimising the damage. You might liken it to a diver who, having made his descent into deep water, needs to control the speed at which he surfaces to avoid decompression sickness.

Trees, of course, have bragging rights on seniority and many don't reach maturity until their dotage in terms of human years. Strong, silent and resilient – they do more for the health of the planet than you could ever imagine.

TREES: Carbon Capture, Congestion and Pollution

The stamina and diversity of trees never ceases to amaze, motoring at full speed all year long and surviving on just a short winter vacation before having to do it all again. Living longer and growing larger than almost any other free-standing organism on Earth, their stoicism and resilience is legendary. It is estimated there are more than 80,000 tree species on Earth and so immersed are they in our culture, we find reference to them just about everywhere. Fables and poems have been written about them, philosophers have pondered their magnificence, and acacia, almond, cedar, fig and pine turn up in the Bible, reinforcing our connection with nature's titans since ancient times.

Topping the honours list for their longevity, trees have offered faithful service for aeons, from igniting Neanderthal campfires to being fashioned into chariots and wheels, furniture and cuckoo clocks. Their timber has provided millions of homes, books, pencils, paper and medicines for generations, supplying leafy shade and charcoal for our barbecues, while building and warming our homes. In the company of trees, you are in the presence of greatness.

Big as well as beautiful, trees make miracles look simple. Pure alchemy. Softly breathing in and out, pulling carbon dioxide out of the air and emitting a miracle end product – oxygen. A single mature tree can produce enough daily oxygen for four people as well as absorbing and reducing noise and environmental pollution. Heralded as being the best natural carbon-capture technology in the world, trees remove carbon dioxide from the air, so it comes as no surprise that our woodlands and forests are massive carbon-capture chambers. The world's forests provide a carbon sink that absorbs a net 7.6 billion metric tonnes of carbon dioxide annually, locking up damaging carbon for centuries – and if that doesn't count as a major miracle, I can't think what does.

Being born in different places, throughout the ages, trees have adapted their external and internal functions and shape to survive, fine-tuning coping strategies to suit their particular surroundings. Some are evergreen, retaining their leaves year round; others shed their leaves as winter approaches. Foliage varies hugely in shape, size and texture, yet roughly only 5 per cent of a tree is comprised of leaves, which prevents them losing too much moisture, and as leaves fall to the ground and rot, they provide food for insects, enrich our soils and provide handy building materials for both humans and nesting birds.

Across British woodlands, mature trees trap up 213 million tonnes of carbon – and it's a case of age over beauty, since ancient woodlands can store more carbon than younger ones. A young wood is estimated to capture 400 tonnes per hectare through their leaves, roots and soil – chickenfeed compared to the United States, where forests capture over 800 million tonnes of carbon annually. The American essayist and poet Ralph Waldo Emerson summed things up rather nicely, reflecting on their essence: 'The creation of a thousand forests lies within one acorn.'

Unfortunately, this green and pleasant land of ours has the dubious accolade of being one of the least-wooded countries in Europe. Just 13 per cent of the UK is forested, so we need to protect the trees we already have and plant many more to maximise their remarkable ability to purify our air – and variety in the trees we plant is as important as quantity because it encourages biodiversity. To reach the UK's carbon neutral target, it's estimated an overwhelming 1.5 billion trees need to be planted by the year 2050. Easier said than done, when restrictive factors such as land scarcity, together with tree plantations that have either peaked in their ability to store carbon or have toppled through old age makes accurate assessments of the numbers needed difficult to pinpoint. The United Nations estimates that there are currently about 4 billion hectares of forest on Earth right now, so to put that in plain English, the Massachusetts Institute of Technology provides this explanation:

> It's tempting to think that a back-of-the-envelope calculation can deliver a useful answer. For example, in 2021, the U.S. emitted about 5.6 billion tons of CO2 and other greenhouse gases. A hectare of trees, meanwhile, might hold around 50 tons of carbon, equivalent to around 180 tons of CO2 in the atmosphere. At this rate, it would take a little over 30 million hectares of trees to account for one year of American emissions – or a forest roughly the size of New Mexico.

Factoring in that thousands of trees are burned for fuel, and in the case of wildfires like the recent blazes in California and southern Europe, trees release all the carbon dioxide they have been sequestering, it makes as much sense to preserve the trees we already have as to plant new ones.

Trees have other hidden virtues, too. Anyone would think that pollution, traffic noise and traffic-calming measures were modern-day challenges. The good citizens of ancient Rome were so crazed by the incessant din and daily chaos of the city that they looked to Julius Caesar to tackle the nuisance. Imagine chariots blocking narrow streets, drivers trading lively insults and punches, horsemen tramping dust onto clean washing hung out by cursing housewives or kids dodging the relentless clatter of wooden wheels over cobbles. The problem of noise and air pollution is still with us after all these years, as those living in busy cities, near airports or in heavily trafficked areas will attest. Anyone with an ounce of common sense realises it's always more effective to reduce noise at source, resorting to fences and physical barriers to deaden noise, but when that isn't a viable option, trees and shrubs come to our rescue. My lanscape clients are always amazed when I suggest planting greenery as a method of reducing noise or air pollution. Can plants do that they ask? You bet they can.

With the unique ability to respond to external chemical stimuli, we commonly harness their sensory capacities to improve our everyday surroundings. Topping the list of sound silencers, larch was deemed the most effective muffler, despite the fact it's the only European conifer to lose its needles in winter – and that's down to the fact it's not always the foliage that's the most noise absorbent. Sure, sound gets scattered across all parts of a tree, but it's the branches and thick bark that proves far more effective in noise reduction. Evergreen trees, oak, eucalyptus and yew, as well as larger evergreen shrubs such as holly and laurel, can reduce noise by several decibels, and even though broad-leaved trees shed their leaves in winter, inclining us to believe they don't offer effective sound insulation, it's worth remembering that deciduous trees provide softer benefits that aren't as quantifiable.

Providing cooling shade, trees steadily filter out airborne pollutants and capture acids, dust, metals and organic chemicals. Known as particulate matter, which can easily enter the human respiratory system, it's estimated that nearly 9 million deaths from lung or respiratory illness could be prevented worldwide if trees were more widely planted in urban areas to absorb nasties from the air. Paris has recently taken the decision to remove half of its 140,000 on-street car parking spaces as it seeks to make the air cleaner and the city greener and more people friendly.

Living museums of natural history, both in urban and rural landscapes, trees provide shelter for countless small mammals, bats, birds, butterflies, moths and other insects, while offering us the potential for escape. Other soft skills include striking up priceless friendships with beneficial fungi to promote soil health, backing up existing flood defences by reducing the volume of water entering our watercourses, preventing burst banks and making our homes less vulnerable to flooding. And as if that wasn't enough ...

FOLIAGE: Trees and Leaves

Robin Hood (*Robyn Hode* in ye olde English) knew the value of our rugged friends as he scampered around the ancient oak forest of Sherwood with his merry outlaws, dispensing largesse and retribution. It's believed the most famously occupied, some say by Robin himself, was the Major Oak, thought to be at least 800 years old and with a canopy extending to some 28m, weighing a hefty 23 tonnes and still in fine fettle today. You'll have to admire it from a distance nowadays as it's fenced off, and although I don't doubt the veracity of Robin's jolly men in tights, I can't think a fence would have dissuaded them. Anyhow, surely a more rugged hosiery was the order of the day. Horror! Can you imagine all that inconvenient snagging?

Talking of ancients, the ferocious-sounding dragon blood tree that famously weeps blood-red resin when cut or wounded has unfortunately achieved celebrity through threat of extinction. Sadly, wild populations are in decline. A mature wild tree still clings precariously to a scrubby slope in Madeira, while two of its near neighbours have been toppled by storms and yet another dived into the sea. Adverse conditions, climate change, overgrazing and human activity have taken their toll, prompting conservationists to step in and monitor their numbers. Thankfully, practical measures are being taken by planting saplings, protecting existing plants from overgrazing and weeding out invasive species in trying to halt their decline.

Native to the Azores, Canary Islands, Madeira, Morocco and Yemen, looking up into a dragon tree's curious canopy, you can't help thinking it resembles a giant umbrella. A sturdy grey-white trunk radiates strong spokes, each topped with shorn blue-green daggers that shade it from the sun's heat. Living for up to 100 years and reaching 40ft in height and girth in maturity, according to Greek mythology these gnarly ancients sprouted from the blood of a 100-headed dragon slain by Hercules. Historically, tribal nomads drained its sap for paint and its resin is commonly used in cosmetics and remedies.

But it is the service these giants offer in local areas that makes their loss both a potential natural and human calamity. Because, unlike the majority of trees that draw water from the soil to their leaves, dragon's blood possess an extraordinary ability to move water in both directions, which accounts for their survival in arid regions as well as cloudy, wet areas. Those perfectly engineered umbrellas snatch moisture from clouds and send it down to their roots, bringing huge amounts of water into local habitats and sustaining human activity, neighbouring plants, reptiles and wildlife. Remarkably, a single tree can force more water into the soil than the local habitat captures in yearly rainfall, so if one tree is lost, hundreds of gallons of water are also lost with unthinkable consequences. That's not to say you can't buy the mini-houseplants, but they are dwarfed by the impressive infrastructure of a dragon tree growing in the wild.

Let's talk about foliage. Because tree leaves aren't green at all. With the arrival of spring, trees start making the most of the warmer, brighter days, putting all their effort into making chlorophyll, a photosynthetic pigment responsible for those infinitesimal shades clothing our trees from spring into autumn. Found in abundance, this pale green pigment absorbs light at the blue/red end of the spectrum – the greater the blue light leaves absorb, the darker the greenery.

Chlorophyll conceals a tree's true nature, which is a natural palette of reds, oranges and golds that are dramatically unmasked with the arrival of autumn. Enticements on travel websites, postcards, TV and travelogues offer us the chance to enjoy this global unveiling from the Catskills to Japan, Finland, Scotland and Vermont, and depending on the weather, autumn displays can vary hugely.

Even though chlorophyll manufacture has pretty much halted, on gorgeous sunny autumn days limited photosynthesis still occurs with more anthocyanin than green pigment being produced, resulting in those dramatic bonfire hues. Which explains why in some years autumn colour can be glorious and in others it can all be a bit dreary. Too cold and with low chlorophyll reserves, means pigment breaks down, turning leaves dun brown rather than those seductive colourful accents.

While tourists are busy snapping a stunning tapestry of mellow autumn golds and ruby reds on their mobile phones, tired from standing about all year in all weathers, trees start slowing down for a well-earned rest. Sensing summer's end as the days grow shorter, and struggling to get enough sunlight to make food, deciduous trees stop producing chlorophyll and begin to shut down, revealing their true colours. Forced to drop their leaves to survive winter – auxin (that handy hormone again), which has kept their leaves attached until now, dwindles, fracturing the bond between leaf and branch in a process called abscission and tosses them to their deaths. With no canopy to support, our green generals can rest easy and exit centre stage, weathering winter storms, gales, ice, hail and snow until life begins anew.

Marking the seasons of the year with pinpoint accuracy, it's hard not to be awed by a tree's silent machinations. It's no revelation that fallen leaves play their part in the wondrous cycle of life, rotting and returning nutrients to the soil. Forests and woodlands offer safe, reassuring spaces where we can reconnect with nature, which explains why many therapists and counsellors have abandoned office-based sessions in favour of the great outdoors. And should you wish to get better acquainted with trees, there's nothing like a woodland walk for a digital detox. The Japanese culture of 'forest bathing' encourages us to leave technology at home and walk in woodlands, taking the time to notice nature all around us. If you're short of a nearby forest, any wooded environment will do to consciously connect with the natural landscape. Sun filtering through canopies, chiming birdsong and gentle winds stir the boughs. In a woodland glade, the ground is soft and mossy underfoot, the air is fresh and clean and modern life takes a back seat for a moment, letting us slow down, relax and breathe. We could do no better than heed American evangelist Robert Schuller's advice, 'Never cut down a tree in wintertime. Never make a negative decision at a low time. Never make your most important decisions when you are in the worst of moods. Wait. Be patient. The storm will pass. The spring will come.'

But of course, trees know this.

CONIFERS: Evergreen Elders

Follow me back to my childhood where a towering Wellingtonia presided over the lower school lawn, with summer afternoons spent sprawled under its cool shade, squinting up at a cloudless sky through its thickly clothed branches.

Giant redwoods, with their spongy, rusty bark, provide an excellent pastime for kids armed with penknives. Any number of names were etched into its trunk, with hours spent in rapt concentration, pink tongues peeping through milky teeth as we each carved our indelible rite of passage. Redwoods are impressive skyscrapers with a heritage that can be traced back to Jurassic times, when they developed 12in-thick armoured bark to protect themselves from forest fires, pests and rot. Wellingtonia also have the largest mass volume in the plant kingdom. *W.* 'General Sherman', found in the Sequoia National Park in California, is testament to the fact. At roughly 2,000 years old, with a dizzying height of 83m, at ground level its trunk diameter measures a staggering 31m plus, endowing it with the accolade of being the world's biggest tree. Truly the gentle giants of our forests, redwoods purify the air by storing carbon, shelter small animals and birds, and are humble enough to realise even a mighty army can't stand alone. Often spreading their roots to distances of over 100ft, by entwining with neighbouring roots for stability, they generously share essential nutrients while supporting over 200 different plant species.

Although not flowering in the accepted sense, coniferous evergreens ensure their survival through very different means to deciduous trees, producing cones instead of blooms. I'm often asked the difference between firs and spruce, and it's pretty straightforward. Fir cones are upright and spruce cones tend to dangle, while their needles are attached in slightly different formations. Typically, conifers have fine, modified leaves because needles are less prone to water loss than larger, broad-leafed species. Needled foliage does the same job in oxygen production as any broad-leaved specimen but there are lots of needles, right? All beavering away, collecting sunlight in their millions and making precious oxygen.

At a glance, you might think needles are all the same, yet they differ in shape and size from species to species. Pine needles come in bunches of two, three or five, whereas the elegant smoke-grey Colorado spruce has almost square-shaped needles, clothed in evergreen greatcoats of such luxury designed to keep this staunch brigade toasty. Needles are practical, too. You only have to look at snow-prone regions or areas of high altitude to recognise that snow slides more easily off a conifer's sloping shoulders, hence the reason many are domed, pyramid or cone-shaped. That clever tapering design also helps maximise light absorption in low light levels, ensuring foliage sited lower down isn't overlooked.

Since trees can't flee or protect themselves from forest fires or escape wintry blasts, that's where tough bark comes in handy – a durable waterproof shield offering insulation and protection from the worst weather and rebutting insect invasion or damage. Scandinavian Norwegian spruce has long since settled its differences with a hostile climate. Clad in thick scaly bark to insulate it from the cold, its waxy-needled foliage minimises water loss and its armour-rigged pine cones protect its seed in the harshest of winters.

Interestingly, in tropical regions tree bark tends to be thin and slippery, repelling other plants from growing on its surface. Being at low risk from water evaporation in humid rainforests, these trees have nothing to fear from frost, snow or ice, making thick overcoats unnecessary. On the other hand, broad-leafed trees have larger leaf areas, which are easily able to absorb light via their wide crowns even when sunlight is eclipsed by cloud, enabling them to cope with degrees of shade. Additionally, their leaves trap warm air, offering some insulation, and while bark doesn't stop wild bears using their trunks as scratching posts, it helps retain moisture and resist air penetration that might otherwise damage living cells.

As with roses and a great number of shrubs, some trees have developed thorns and spines to protect their leaves from being eaten and chemicals such as tannins in tree bark also contribute to their survival. Alder, birch, oak and willow are jam-packed with the stuff, helping them repel bacteria and harmful herbivores.

Another epic strategy is the ability to self-heal. Trees produce gum and resins that block or help heal infections. If you've ever spotted a hard, glossy toffee-like resin on ornamental cherries, it's a sure sign a tree is seriously sickening and trying to heal itself. Fortunately, trees are brilliant revivalists, which is down to the fact they produce cambium. That thickening tissue again, sandwiched between xylem and phloem, that forms sapwood, a softer outer layer between the inner heartwood and bark. Sapwood, of course, is where you'll find those medullary rings, allowing us to estimate a tree's age with a fair degree of accuracy. Similar to human stem cells, trees have different cells serving different functions and cambium is extraordinary insofar as it can assume any function it likes, although it is primarily focused on secondary growth of roots and stems, thus increasing the girth in woody plants.

I'm frequently reminded of their restorative power by an avenue of tall limes in a neighbouring garden that undergo heavy pollarding every few years in late summer, once their leafy crowns have thrown the house into shade. Up march the tree surgeons wielding their screaming chainsaws, reducing them to nude knobbly stumps. Give it a few weeks, and bang on target, the heads are speckled green again. When heavily pruned or injured, urgent signals are telegraphed to HQ and trees grind into survival mode, kick-starting growth hormones to produce new leaves – and damn quickly. Interestingly, some trees can cope with being shorn within inches of their lives. Yew, for example, can be reduced to almost ground level and reshoot. Others just can't cope with the trauma. Leylandii that are cut back drastically to old wood are done for – the subsequent rusty bald spots never regain their greenery.

PART FIVE

GAME-CHANGERS: PLANTS THAT SHAPED HUMAN HISTORY

Man selects only for his own good: Nature only for that of the being which she tends.

– Charles Darwin, *The Origin of the Species*

Certain plants have profoundly shaped human history, influencing economies, cultures and even conflicts. Cotton fuelled the global textile industry and drove historical trade networks, along with slavery and industrialisation. Cannabis has been used for fibre, medicine and recreation, and sparked legal debates and cultural movements across centuries. Plants like agave, barley and grapes gave rise to alcoholic beverages that have crowned and toppled kings, boosted trade and become central to rituals, economies and social life. Plants are more than just crops or ornament – the evolution of plants and society are deeply intertwined, catalysts that continue to influence human history.

HEROES: Contraception, Cannabis and Opium

While Culpeper was busy building his reputation in sixteenth-century London, millions of people around the world living in mosquito-infested regions were falling prey to a deadly disease. At a time when Portuguese traders, rapidly succeeded by European powers, attempted to establish colonial outposts on the African continent, they were repeatedly thwarted in their ambitions by hostile indigenous people and tropical diseases, including yellow fever and malaria. But malaria, the 'killing fever', had the greatest impact in Africa – no other continent, past or present, has been so badly afflicted; by the eighteenth century, the rampancy of the disease across Central and Western Africa gave rise to the region's epitaph, 'white man's grave'.

In the early days, malaria was commonly treated by blood letting using leeches or inducing vomiting. The first known cure for this killer disease was allegedly found in the forests of the Andes in the early 1600s. An indigenous Indian stopped to drink at a stagnant pool that was surrounded by *Cinchona,* or fever trees, from which quinine is derived. Noticing the water tasted bitter, he worried that the water was poisoned, but hours later his fever unexpectedly abated, and he shared his recovery with the locals.

Europeans share a divergent account – claiming that a Jesuit priest combined cloves, quinine tree bark and rose leaf syrup, which saved the life of the Viceroy of Peru's wife. It's alleged it was *actually* Mrs Viceroy who introduced quinine to Europe, but it took 200 years and two French chemists in 1820 to finally isolate the quinine compounds that acted as the launch pad for the most effective treatment of malaria ever known. It's a pretty marvellous reassurance, as you sit back and sip that perfect G&T on a warm summer's evening, that quinine is regarded as one of the most important breakthroughs in medical history – one that has saved millions of lives.

Deserving of a place in our hero's gallery is willow. The earliest use of willow bark is documented around 1500 BC, when it was commonly used by early healers for fever and pain relief. Containing salicylic acid, willow bark protects trees against insects and fungal diseases. However, many centuries later, a German chemist, Felix Hoffman, working for the German pharmaceutical company Bayer, synthesised a duo of Jekyll and Hyde drugs (although not at the same time), heroin and salicylic acid. By 1874, Hoffman had distilled salicylic acid into a stable form, giving rise to one of the most widely used drugs of all time – aspirin. An interesting historical footnote reveals that Bayer, realising they had a potential money-spinner on their hands, marketed the drug globally, securing a patent in the United States but not elsewhere. It transpired that acetylsalicylic acid had been previously synthesised, but not in a pure, stable form, which brought Bayer's monopoly on aspirin to a sticky end when its American factories were sold in 1919 as part of Germany's war reparations after the Great War. Bayer's misfortune was human gain because suddenly aspirin was available as a cheap over-the-counter drug worldwide.

As well known is opium, a highly addictive compound derived from *Papaver somniferum,* which translates to 'sleep inducing', and not to be confused with ornamental poppies or the delightful red corn poppies we met earlier. Opium's chequered history makes for interesting reading. An elixir used in early cultures for healing, curses and portents, one of the earliest references to opium cultivation dates to the ancient Sumerians of Mesopotamia, around 3400 BC. Before long, the so-named 'Joy plant' travelled to Egypt, Italy and China, before finding its way to America and Europe. By the eighteenth century, opium seeds had been distilled into a wonder drug – morphine. Offered to the sick and wounded of the American Civil War to relieve their pain, it's estimated the Union Army alone dispensed 10 million opium pills to its soldiers. With the introduction of the hypodermic needle in 1856, increasingly physicians started over-prescribing morphine to treat all manner of chronic illnesses and afflictions, knowing there was little it couldn't relieve. Unfortunately, recovering soldiers treated with this new miracle drug during combat often returned home to find themselves heavily addicted to their medication as widespread prescriptions precipitated a hideous own goal – America's early opiate addiction.

For the poppy-obsessed gardeners out there – a bit of gardening wisdom always comes in handy. If you grow ornamental poppies from seed, you have to get the timing right. Making the move from greenhouse to garden when your plantlet's roots aren't too long and before the first leaves have fully formed hugely improves the chances of successful transplantation – just in case you were wondering.

High up on the leader board for social impact are wild yams. It seems scarcely believable in modern times that historically the only surefire way of avoiding pregnancy was through abstinence or women pleading headaches. Naturally, crude barrier contraceptives have been around for centuries, probably more useful in countering sexually transmitted disease than pregnancy prevention. The first noted use of condoms occurred in the sixteenth century, with a small linen sheath drenched in a saline-and-herb solution that sought to protect wearers against syphilis, devised by an Italian anatomist Gabriele Fallopia (yup, that's where the term fallopian tubes hails from). Other barrier devices of the time included sheaths made from tortoiseshell or animal gut and, again, these were more likely employed against STD infections.

In medieval times, women of all classes would insert herbal pouches or use douches made from wild plants in the hope of avoiding pregnancy or aborting an unborn foetus. Poor and working-class women must have dreaded the thought of yet another hungry mouth to feed or being warned by priests to *be more careful* next time. In a culture that perceived sex as sinful, the stentorian finger-wagging of the prevailing male clergy only cemented a widespread sexual ignorance, creating a perfect environment for sexual abuse.

So, who would have thought that wild yam, a climbing vine with appealing heart-shaped leaves and found across Asia and Mexico, would deliver a seismic change for both married and single women across the globe? Wild yams contain diosgenin, a naturally occurring compound that was chemically converted in the 1960s to produce the world's first contraceptive pill. Having conducted a straw poll just for fun as I was writing this book, results showed that many modern women had no idea that the luxury of modern contraception is all down to this humble vine – well, now you do!

While many welcomed this plant-based miracle, in developing countries (and even in modern Ireland), it was inaccessible for millions through poor government provision or on religious grounds. Irish women were forbidden any use of contraception, influenced by the Offences against Persons Act of 1861, and were forced to rely on home remedies, abstinence or natural methods. As late as 1935, advertising or selling contraceptives was illegal and a further government Act in 1946 banning any publications or books on the subject compounded the issue, ensuring women and young girls remained ignorant of the rudiments of sex, let alone pregnancy prevention. Naturally, Ireland being a Roman Catholic preserve meant abortion wasn't an option, either. Outlawed by a referendum in 1983 that recognised the rights of the unborn child, it made the act of even seeking advice on termination outside the state illegal. News footage of the era shows women waving condoms aloft as they disembarked from what was dubbed the 'Condom train', when they travelled to big cities like Belfast where no such barriers existed to procure contraceptive 'rubbers', 'Johnnies' or 'French letters'. Unbelievably, it wasn't until the Irish government defied the Catholic Church in 1985 that the contraceptive pill became more widely available.

An equally illuminating discovery of the 1960s prompts another landmark episode. American medical researchers found that bark from Pacific yew contained a revelatory ingredient, taxol, which could be employed to reduce cancer cells. Estimating that massive tree production would require three mature trees per patient, the US government fortunately found high levels of taxol in European yew and modern methods have allowed extraction without the need for widespread deforestation and tree felling. Taxol still contributes to the treatment of breast, lung and ovarian cancer, and is used in anti-cancer drugs today, with numerous physic, botanical and historical gardens donating their annual yew clippings for medical use. In fact, I can barely think of anything affecting our daily lives that isn't connected to plants in some way: their green fingerprints are everywhere.

TRADE: Cotton, Cocaine and Tobacco

COTTON

Deserving of a place in our hero's gallery for sheer versatility is cotton, one of the most durable plants on Earth. A close relation to hibiscus, cotton is used more than almost any other natural fibre, providing endless comforts from underwear, T-shirts and hoodies to fertiliser and everything in between. It's not exactly known how long we've been cultivating cotton; the ancient Mayans and Aztecs were enormously adept at woven textiles and are thought to have been among the first to fashion clothing from it, and even in ancient days, it was available in subtle shades with the use of vegetable dyes. Pakistan was growing, spinning and weaving cotton on home-made looms from around 3000 BC, long before the Egyptians hijacked the reputation for cotton excellence – although, I daresay the modern preoccupation with thread counts might well have left them scratching their heads.

As versatility goes, cotton is a top scorer with almost every part of the plant put to good use. Like any flowering plant, seeds (called boils) follow the flowers, and the ripened seed splits open revealing those trademark fluffy inners. With countless uses, from bedding, linens and paper products to plastics and even explosives, additionally, cotton seed is crushed for use in livestock feed and fertilisers. Linters, the fluff left on the seeds, makes our cosmetic cotton pads and what is left of the plant, stems, stalks and leaves, is recycled through composting and returned to the soil.

Alongside the smell of freshly mown grass, there's nothing quite like slipping between freshly laundered sheets with that distinctive cotton-fresh scent. But its reputation hasn't always been snow white. In what is known as the 'middle passage', millions of Africans – men, women and children – were kidnapped from their homelands to underpin cotton production in the West Indies and the Americas, contributing to the largest forced migration in human history.

On graduating from Yale in 1792, Massachusetts-born Eli Witney couldn't have imagined that a chance trip to a friend's plantation in Georgia would transform the US cotton industry. Eli, witnessing the difficulty of harvesting cotton by hand, came up with a solution – the cotton gin. Patented in 1794, he created a device that could strip cotton from seed so rapidly that it could easily replace fifty enslaved cotton pickers. You might think the device was a good thing for the pickers who worked the fields, growing and picking cotton from dawn till dusk, ill-treated, malnourished and deprived of their freedom. However, fuelled by the Industrial Revolution in both Britain and the United States, by the mid-1800s, cotton had eclipsed all previous commercial crops, becoming the mainstay crop of the Southern states and the United States' chief export. Its commercial success, while making millionaires of many, brought untold misery to others and as the cotton boom raged – one that even to the present day hasn't subsided – instead of requiring fewer workers, demand for American cotton continued to outstrip supply, providing enslaved workers with no relief from their back-breaking toil.

Cotton remains one of the biggest industries today, with a world trade value of approximately $63.9 billion, with the United States retaining poll position with exports worth around $10.9 billion. If this period of history interests you, I would urge you to read a poignant, yet courageous account of one man's epic journey from his native Guinea as he struggled through enslavement to become a free man. It's entitled *The Interesting Narrative of the Life of Olaudah Equiano, or Gustavas Vassa, the African* and is an uplifting tale of resilience and heroism, written by himself and free to read online at www.gutenberg.org.

COCAINE

Which of us wouldn't be buzzing with energy if we drank wine spiked with coca leaf every day? Some plants rarely grown in a garden have gained global recognition. Coca, one of the oldest cultivated crops, was entwined in the daily lives of native Andean tribes of South America, who were reliant on it as a dietary supplement packed with vital vitamins and minerals. As well as being used for medicine and spiritual rituals, it was also used as an energy booster to take the edge off fatigue, hunger and thirst. Leaves were brewed for teas and tinctures while dried leaves were mixed with ash and chewed – a stimulant still used by Bolivian long-distance truck drivers, miners and manual workers today.

Coca plants slipped from being a daily sustenance of indigenous peoples when they attracted the attention of Spanish conquistadors in the early 1500s. Conquering colonisers regarded the habit of chewing coca leaves as ungodly, but such was the demand for the frost-sensitive crop that could be only grown at low altitude that it was energetically bartered in exchange for foodstuffs.

Marriages, births and even a woman's first menstruation were gifted with coca leaves. After a failed attempt at banning its use, which collided with a deeply ingrained resistance among local people, the King of Spain relented and decided to tax coca leaves instead. (Some things never change.) By now, coca was a lucrative commodity worth 1 million pesos annually and it began to be traded globally as plantation owners hijacked its production. With the arrival of the silver rush across the United States, Canada and Mexico in the 1800s, mine owners realised that workers could labour longer and harder if they received regular coca leaves in payment.

The history of the coca plant was to take a darker turn when a crop that had sustained indigenous peoples for centuries was co-opted by South America's drug barons. Although there are approximately 270 known species of coca, only two – *Erythroxylum novogranatense* and *Erythroxylum coca* – contain the potent alkaloids essential for producing cocaine, a drug that the US government outlawed in the early 1900s.

From drug lords to brand leaders – in the late eighteenth century, Vin Mariani was lauded across the world. A potent tonic blending red wine and coca derived from different species of coca plants indigenous to South America and developed by an entrepreneurial French chemist, Angelo Francois Mariani, Vin Mariani was the global leader of its kind in its day. Touted as a cure-all for everything from chills to consumption, fever, anxiety, depression and malaria, with a catchy slogan claiming it 'fortified and refreshed body and brain', it was served widely as a popular cocktail blended with bitters, vermouth and a twist of lemon. It was imbibed enthusiastically by politicians, housewives, movie stars and even renowned personages including Sherlock Holmes, Sigmund Freud and the Pope. Its popularity remained undented by the introduction of the US Congress Pure Food and Drug Act of 1908, which insisted that ingredients on consumer products should be clearly labelled. Inevitably, the onset of US Prohibition saw sales slump, and Vin Mariani's reinvention came in a reformulated carbonated tonic – replacing cocaine with higher doses of caffeine and sugar. Despite the removal of its original ingredient, Mariana's new carbonated drink attracted an ever-increasing following: Coca-Cola was born and still remains the world's best-selling soft beverage to this day.

TOBACCO

On page 259 of Linnaeus' *Species Plantarum*, we find an entry for *Nicotiana tabacum*, a tropical herb with attractive white flowers and another epic raw material. Tobacco cultivation easily dates to 3000 BC, when it was grown by the Bolivians for religious and medicinal purposes. Leaves were chewed or smoked as a stimulant to help them connect to the spirit world.

Ancient civilisations like the Mayans were smoking tobacco as early as 5000 BC, illustrated by artefacts depicting them smoking rolled tobacco leaves and pipes. Believing that clouds were made from smoke, they made offerings to the Earth Gods, asking for favourable weather and bountiful harvests, and by the time Christopher Columbus landed in South America around the mid-fifteenth century, tobacco leaves had become common currency.

Bartered by Portuguese, Spanish and, later, English mariners, it was John Rolfe who first began cultivating tobacco in the early 1600s in one of the New World's earliest English settlements in Virginia, with the intriguing historical footnote that he married the daughter of a Native American chief – none other than Pocahontas. By the 1800s, cigarette-rolling machines had been invented and by the 1900s, heavy-lidded movie stars popularised social smoking in moody black and white cinema as billions of 'coffin nails' were sold all over the world, making it a key cash crop in the US economy.

DRUGS AND NARCOTICS: Poison and Addiction

POISON

Let's meet angel's trumpets, members of the nightshade clan – and knowing the family link might give you a clue to its super power. Brugmansia are stunning South American shrubs and trees, hung with huge pendulous blazons in dreamy pastel shades. Also used as a recreational drug, it contains a particular compound associated with LSD, which historically featured largely in shamanic rituals or was used medicinally to alleviate pain and healing.

Brugmansia are magical in more ways than one – all parts are toxic and hallucinogenic, a feature I haplessly harnessed for my own ends. In growing frustration at having my immaculately clipped box hedge continually decimated by a neighbour's goats, I put my horticultural expertise to good use. Or so I thought. I planted a barricade of them, thinking to dissuade the blighters. Forgive me, but most sensible animals instinctively know which plants are harmful and those that are best avoided, but it would appear this particular herd hadn't read the plant encyclopaedia that day. By the following morning my trumpets had been razed to the ground, and none the worse for wear for their night of debauchery, those wretched goats continue to graze on my greenery to this day.

To my knowledge, goats at least have the good sense to leave *Strychnos nux-vomica* alone. A harmless-looking deciduous tree found growing along Sri Lanka's riverbanks, you'd never guess at its capacity for mayhem. After the white flowers fade, the tree produces bitter, inedible, apple-sized fruits tucked among bright green leaves. Inside each fruit, the seeds harbour strychnine, a potent toxin that wards off hungry animals and disease. But once ground down into a white, odourless, bitter powder, it proves lethal in the right doses.

Commonly used in rat poison, and easily obtained from any apothecary of the day, the British crime writer Agatha Christie wasn't averse to killing off her victims with a deadly dose of poison. Cantankerous grand dames and young widows suffering hysteria and palpitations were famously treated with small doses of strychnine, but it was on the foggy streets of Victorian London that Dr Neill Cream catapulted strychnine into the hall of fame. Cut-throats, villains and pickpockets abounded in the rotten borough of Lambeth, not the sort of place that lone women lingered at night unless they were ladies of the night. And they could be forgiven for believing that a fashionably attired gentleman in his silk top hat, shadowing thick sideburns and an impeccably trimmed moustache was no different from any other wealthy punter, willing to pay good money for services rendered.

Imagine a rat, its fur slicked wet, black eyes shining in the darkness. Pausing on its haunches, its whiskers twitch before scuttling off to tend to its young, its belly fat from a good night's scavenging.

Within seconds it was dead. Its symptoms identical to those of Cream's victims. Ten women encountered the affluent charms of Dr Neill Cream and within hours of meeting him were found dead. On trial at the Old Bailey in November 1892, the 'Lambeth Poisoner', as he became known, was found guilty of murder and sentenced to hang, rendering Cream, along with Jack the Ripper, one of the most notorious serial killers of the Victorian era.

ADDICTION

Let's touch lightly on hemp and cannabis, related species originating thousands of years ago in central Asia around 2700 BC and with the proud heritage of being one of humanity's earliest cultivated crops, relied on for cloth, rope and oil long before we began rolling them into spliffs. The father of Chinese medicine, Emperor Shennong, was one of the first to document hemp's efficiency for the treatment of poor memory, malaria and rheumatism. From here, hemp's prospects diverged, spreading to India and becoming central to Ayurvedic medicine before heading out into the wider world.

Onward to Ancient Greece and Rome before hitting the Middle East and North Africa when the Arab world used hashish, a concentrated form of cannabis, for medicine and psychoactive purposes. By the 1500s, this pungent herb had found its way to Virginia, United States, where its usefulness was cemented in making rope, sails, shipbuilding and paper. Yet, it was only in the early 1900s that its reputation was debased by its close cousin, *Cannabis sativa* – pot or marijuana – which had long been used in Mexican culture and by a growing influx of Mexican immigrants into the United States. It triggered anti-cannabis sentiments, linking its use to criminal behaviour and exacerbating racism and xenophobia.

By 1937, the United States had effectively criminalised it with the introduction of the Marihuana Tax Act. However, rising above its illegal status from 2010, many US states have since legalised its use, and more recently, cannabis derivatives have gained a more savoury reputation, increasingly used in CBT (cannabitriol) products, chemotherapy, epilepsy and multiple sclerosis.

Equally important to our culture are botanical beverages. How do you like yours? Shaken or stirred?

BOOZY BOTANICALS: Intemperance and Sobriety

We'd be missing a trick if we omitted the huge contribution plants have made to human libation. Anyone who has ever read H.E. Bates' *The Darling Buds of May*, set in the Garden of England, can almost smell the beery waft emanating from its pages. Where would beer be without hops? Wine without grapes? Or whisky without rye?

From earliest civilisations, humans have concocted liquor using plant sugars and transformed them into boozy beverages. At Cain's biblical wedding feast, Jesus transformed cheap, vinegary wine into a more palatable vintage – a gesture, no doubt, for which the embarrassed host was supremely grateful and proving the early Greeks and Romans were not averse to the odd tipple.

'*In vino veritas*' – 'In wine, there is truth'. So said the Greek philosopher, Pliny the Elder, who obviously had some experience of the local plonk, and if the tradition of grape treading conjures up images of hairy-toed peasants stomping their grapes (seedless or otherwise) in a charming hamlet somewhere in France, it was the early Greeks who initially cornered the market. History tells us that another Greek philosopher, Socrates (who, incidentally, never set down his wisdom in written form yet he still remains one of the most celebrated sages of all time) was not averse to the occasional tipple:

> So far as drinking is concerned, you have my hearty approval, for wine does of truth moisten the soul and lull our griefs to sleep (and with small cups) we shall ... be brought by its gentle persuasion to a more sportive mood.

Long subscribed to grape stomping, early Greek wines were highly prized and sweetened with honey and herbs, resulting in a wine so sweet that it was seldom drunk neat.

The Romans being Roman soon bumped the Greeks off top slot, producing one of the most expensive white wines of the time, Falernian wine. A single bottle might set you back a year's salary and by outlawing wine production in other regions, Rome retained the monopoly on their prized vintage. If you're a wine buff, you'll be interested to know that Falernian wine is still available today, though at more competitive prices.

Japanese sake has been around since the eighth century, and although we tend to think of monks presiding over a brewing monopoly in the Middle Ages, ale making was equally a woman's business. Known as brewsters or maltsters, nunneries and housewives had a pretty impressive track record in producing homebrews. While their menfolk laboured in the mills and fields, goodwives turned their hands to all kinds of housewifery. Soaking barley grains in water produced enzymes that turned grain starch into sugar, and I like to imagine that quite by accident, a crafty housewife, tiring of a haranguing husband, discovered that adding malt to dry barley enlivened a basic household beverage no end (and presumably numbed her against any masculine carping).

Imagine then, a foreign traveller arriving in London at the start of the eighteenth century, only to discover that although water was everywhere, almost no one drank it. Much of the water supply in cities would have been riddled with filth – animal and human waste, plus other impurities, making it more hazard than refreshment – and it wasn't until the twentieth century that drinking water was reliably sanitised. In the fifteenth century, the arrival of ale, a basic beverage regarded more as a household staple, was safer than drinking polluted water as it took its place alongside mead, milk, cider and perry.

Pastoral scenes of golden fields of rising barley belie ale's supremacy since the Neolithic age. Although the exact origins of ale have been lost in the mists of time, it's likely the oldest known alcoholic beverage and its story probably began alongside early domestic grain farming around 10,000 years ago. Evidence suggests that the first homebrews were rudimentary concoctions, fermented blends of barley and water with added flavourings such as cinnamon, nettles and milk, together with locally grown herbs and other additives that gave each a distinct flavour according to region.

Hops, with a more bitter taste and containing essential preservatives, weren't added until much later, so in Anglo-Saxon times its shelf life would have been short. No doubt explaining the unquenchable thirst of Vikings and common people alike, when the average man would drink a gallon of beer a day, with diluted versions consumed by children.

The titled classes preferred wine. A leatherbound volume by thirteenth-century English philosopher Roger Bacon on display at the Royal College of Physicians in London recognised the perils of excessive consumption, 'If it be over-much guzzled, it will on the contrary do a great deal of harm; for it will darken understanding, ill-affect the brain ... beget shaking of the limbs and blearyeyedness.' (A far nicer term than the modern-day 'hangover', don't you agree?)

It was Scandinavia that first introduced *aquavit*, or 'water of life', to Europe and the wider world around the fifteenth century, the earliest reference for botanically steeped alcohol and most likely brewed for its health benefits at a time when epidemics abounded. Norway used potatoes as the key ingredient, whereas Sweden largely used grains infused with fennel or anise in their blends, with the Celtic translation for *aquavit*, '*uisge beatha*', giving rise to the English word 'whisky'.

Not to be outdone, Ireland was innovating with a boozy beverage of its own. *Poitín* (or Poteen) is probably one of the best-known bootlegged liquors conceived by the Irish under the many miseries imposed on them by English rule. Irish moonshine, originally derived from grains and, increasingly, potatoes, was created when farmers would retain a portion of their crop to ferment and distil. Once the

British government began imposing punitive taxes on distilled beverages, followed by a total ban on *Poitín* in 1661, undeclared stills sprung up in mossy enclaves all over the Emerald Isle. Most standard whiskies are diluted with water to a strength of 46 per cent ABV, whereas home-brewed *Poitín* delivers a whopping 40–90 per cent ABV (and you may be surprised to learn that the Irish government only lifted the ban as recently as 1997). Nowadays, *Poitín* is quite the thing; experiencing a comeback with micro-distillers creating their own unique blends, although it still needs to contain a minimum 40 per cent ABV to qualify as the real thing. Oh – and I'd quite forgotten to mention that long before its final death throes, the century plant, among other species of agave, is of course essential in the making of fiery *Mezcal*.

Vodka's provenance is still hotly debated between Poland and Russia today, but whatever its origins, we know that it was first distilled in cold regions in the Middle Ages using potatoes or rye instead of grapes. It was a harsh brew designed to be swallowed fast for its warming or galvanising effects (or both!) without savouring the flavour. Nothing as sophisticated as distilling occurred back in the day. Vodka was left outside in sub-degree temperatures because water freezes at a higher temperature than alcohol. They would just scrape off the iced water and drink what was left – neat. Naturally, this meant that any locally brewed liquor was full of impurities, so botanical essences, herbs, leaves and fruit were added to mask any unpleasant taste. From the Middle Ages onwards, vodka had become the drink of the masses, cheap to make and delivering the required numbing of the working classes – although, as frequently, it was used as an antidote to pain. By the seventeenth century, distilleries came into being, producing a more refined liquor and it wasn't until the late eighteenth century that vodka was actually cleaned, removing all impurities to resemble the vodkas of today.

Yo, ho, ho, and a bottle of rum! Pirates of the Caribbean always had a swig of grog at the ready, and without a certain swaying South

American grassy perennial, mincing Captain Jack Sparrow would have found 'water, water everywhere and not a drop to drink'. As Europeans colonised the West Indies, they discovered the merits of sugar cane – a plentiful crop that after being turned into molasses could be fermented and distilled to create a potent rum with an alcoholic content of 70 per cent.

Rum was first documented by a visitor to Barbados in 1647, who reported, 'The chief fuddling they make in the island is Rumbullion, alias Kill-Devil ... made of sugar cane distilled, a hot, hellish and terrible liquor.' Rum did more than strengthen a seafarer's mettle. Once mixed with lemon juice and sugar, it deterred scurvy, a killer disease caused by vitamin C deficiency, and quickly became a British Navy tradition from the 1700s, helping relieve sailors' boredom after months at sea. It was as much a health measure as recreational, because fresh water stored on board stagnated quickly, resulting in algae blooming in the wooden casks. Beer proved a viable alternative until it was discovered that in southerly latitudes it spoiled almost as quickly. While whisky, with a higher alcohol content, better survived long voyages but proved an expensive mood-lifter for lowly seamen.

As the demand for sugar increased, plantation owners found themselves with glut crops that exceeded local demand and turned to Royal Navy pursers to dispose of the excess. Being that British Colonies were forbidden from selling any sugar cane-derived alcohol, the Navy pleaded that rum was more necessity than booze and by 1687 it was officially adopted as an essential in a sailor's daily ration, ensuring the continuing well-being of the king's mariners. Our taste for rum was quick to catch on and in what was to become the United States, Dutch settlers in New Amsterdam (or modern-day New York) were quick to get in on the booming market, and the first of many American rum distilleries was founded.

Oddly enough, it was the Dutch King William of Orange who put gin on the map. Between the late seventeenth and early eighteenth centuries – and even today – to be officially classified as gin, it needed to have a close acquaintance with the dull, tart berries of evergreen juniper, common to Europe and most of the Northern Hemisphere. With political and religious tension rife between France and Britain, England's Catholic James II was deposed in favour of his Protestant Dutch successor, William, who, in a canny attempt to ingratiate himself with British landowners at a time when England was experiencing a grain glut, put his knowledge of the grain economy to good use. Under similar conditions, gin production had boosted the grain market in his home territory of Holland, so immediately he legislated for the increased production of gin.

Overnight, production and availability drove the price of gin down and its desirability through the roof, amidst a landscape of failing and falling births, rising prostitution, death and crime rates. As gin fever gripped the nation, 'Mother's Ruin' took the edge off the mind-numbing misery of the working poor, especially in London's slums, when, between 1730 and 1749, 75 per cent of under-5s died: shocking figures largely attributed to negligent mothering. A working woman of the time grumbled, 'We market women are up early and late, and work hard for what we have and if it were not for gin, we should never be able to ... keep body and soul together.'

No fewer than eight Gin Laws were passed to curb this new social evil, driving it underground as bootleg operations sprang up across the country, fuelling a black market for gin-soused citizens. It wasn't until the anti-tippling act (officially known as the Gin Act) came into being in 1751, outlawing gin distillers from selling spirit to unlicensed merchants, that gin consumption significantly decreased.

The gin trend continues unabated today with an ever-increasing focus on infused botanicals from independent distillers. Famously, early-blooming blackthorn, belonging to the rose family (thorns are a family trait), is responsible for those tart, blue-black sloes so favoured in making the perfect G&T.

Phew! After all that heavy drinking, it's almost a relief to meet plants that offer gentler refreshment.

GENTLE REFRESHMENT: Coffee, Tea and Sugar

You're probably thinking that coffee is a South American invention. I know I did. Where else would that rich Blue Mountain roast come from? Yet, it was widely cultivated in Arabic culture for spiritual purposes and its energy-boosting qualities long before it attracted a popular audience. Thought to have been discovered by a lowly Ethiopian shepherd, who noticed his goats were rather livelier than usual after grazing on strings of red berries from a native shrub growing at high altitude, nowadays, two-thirds of the world's coffee crop is grown in Latin America.

In its early years, coffee, with its stimulating qualities, reached Mecca, where religious leaders fought to have it banned, lumping it together with alcohol – a substance forbidden in Islamic culture. Amid claims that coffee provoked radical thinking and political unrest, legislation had little effect once Sufi Muslims began drinking gallons of the stuff during Ramadan to take the edge off fasting and help them stay awake through night prayers. Its popularity elicited coffee houses to spring up on every street corner – typically all-male preserves where, alongside putting the world to rights, the social and religious issues of the day were energetically debated – and it wasn't long before it arrived in the Ottoman Empire, before finding its way into Europe.

Initially viewed with suspicion as a peculiarly Muslim beverage, its popularity in Paris and Vienna invalidated any government moves to restrict public drinking in newly founded coffee houses, which were largely viewed as dens of thievery and sedition. It was only when Pope Clement VIII, having enjoyed a cup of the rich, dark brew, decreed it would be 'improper' to allow Muslims the monopoly of coffee consumption, that coffee gained wide appeal and acceptance. The rest, as they say, is history.

Second to coffee, tea is the world's most popular beverage. History tells us that around 2737 BC, the Chinese Emperor Shen Nung, relaxing under a shady aromatic evergreen, noticed a few leaves drift into a bowl of heated water, infusing it with a refreshing flavour. Truth or myth, all tea, whether black, green, white or yellow, originates from two bushy evergreen shrubs, *Camellia sinensis,* from China, and *Camellia assamica*, from India. Tea trees in the wild can easily attain heights of 15m, although crops nowadays are kept conveniently waist high for commercial production. One of the goods carried by Dutch merchant ships during the seventeenth century, the Chinese had been drinking it for centuries before it made its way to Europe and later New York. Initially a luxury item enjoyed by the upper classes, it was here that its fate converged with the extensive sugar plantations of the West Indies and Cuba.

Established in 1647, wide-scale cultivation of 'white gold' plantations spawned the misery of the millions of Africans enslaved to tend countless hectares of sugar cane by their colonial masters. But cane was never an easy crop. It demanded backbreaking labour, with rows dug and planted by hand and cut the moment they ripened since the canes spoil within days of harvest. Together with the vast amounts of firewood required to fuel great stoves that boiled the sweet extract into granules and molasses, it only ramped up the pressure on enslaved workers. Today, the United States makes about 9 million tons of the white stuff annually, ranking it sixth in global production.

After cereals and rice, sugar cane occupies roughly 26,942,686 hectares of land across the globe, although it was pretty much unheard of outside its native Papua New Guinea some 10,000 years ago, when it was used locally as little more than spice or medicine and likely to be a gritty brown gravel made from drying the cane extract in the sun. By the fifteenth century, Madeira was first to launch large-scale sugar refineries, followed swiftly by the Portuguese, who realised that Brazil's climate was ideal for cultivating canes. Their foresight triggered a lucrative sugar rush, and soon other nations joined the scrum. By the nineteenth century, producers had found new sources such as sugar beet, but even today, our craving for its sweetness makes it one of the world's most in-demand crops – and a key player in the modern obesity crisis.

Despite its restrained reputation, tea has had a surprising impact on history and was one the many ripples that prompted the American Revolution of 1773. Ruled from afar by the British, American colonists grew increasingly dissatisfied at the hefty taxes imposed by Britain on imported foreign goods. Growing unrest and persistent protests forced the British to lift the majority of taxes, yet they stubbornly enforced a tax on tea exclusively shipped by the London-based monopoly, the East India Company. Permitted to dock at Boston Harbor by the presiding pro-British governor, furious colonists boarded British ships lying at anchor at night, dumping the precious cargo overboard. Known as the Boston Tea Party, this galvanised the movement to end British rule.

The American screen siren Lauren Bacall's perception of the British obsession with tea was well observed, 'The British never seem to do anything until they've had a cup of tea, by which time it is too late.' She wasn't wrong. So devoted to the notion that tea makes everything better, in 1942, the British government bought the entire available supply of black tea in Europe to boost the morale of Allied troops – a hefty expenditure that reputedly exceeded the cost of artillery shells and explosives.

Following combat comes rest and relaxation, prompting us to explore the softer virtues of our fine green friends.

PART SIX

NATURE AT WORK AND PLAY

We know what we are, but know not what we may be.

– William Shakespeare, *Hamlet*

Gardening is more than a pastime. It is a quiet collaboration between human hands and living things. Plants offer beauty and food, but their benefits run deeper. In tending them, the rhythm of growing, watering, pruning and harvesting eases stress, boosts health and improves our parenting while renewing our sense of wonder. Whether you have endless acres, a patio or balcony, our outdoor spaces become both sanctuary and sustenance, grounding us in nature's slow, steady work.

COMPANION PLANTING:
Friends with Benefits

'*Partout où la vie vous plante, fleurissez avec grâce*' – 'Wherever life plants you, bloom with grace.' So says an old French proverb, and maintaining harmony in our everyday lives means hanging out with the right crowd. Mixing with like-minded individuals who share our values, sharing happy moments with family and friends who make us laugh, who stick by us through thick and thin, are happy at our successes and sympathetic to our failures. That's the kind of mood you're gunning for in a garden.

Companion planting equals natural camaraderie and, like us, plants thrive with affable companionship. Surrounding plants with a handful of empathetic chums has a positive influence on the garden environment. It improves soil, deters pests, disease and bullying weeds, prevents water evaporation, and in some cases, as in beans and green manures, helps improve soil fertility. There's no happier sight than azure-eyed borage under a nimbus of bees and butterflies, boosting its own and its neighbours' chances of pollination. In fact, any nectar-rich plants grown together to create floral perimeters encourage more pollinators into flower borders, fruit and food crops, whatever they might be. And although we tend to think of companion planting as being a relatively modern idea, it's been happening for years.

Native Americans were early advocates of the mutual appreciation society, growing beans, corn and squash together. Known as 'the Three Sisters' system, they combined different crops, having sussed out that they provided mutual benefit by creating a mini-ecosystem. It works something like this. Grateful for Bean's generosity, eldest sister, let's call her Corn, allows Bean to climb her tall stems, providing support and protection from damaging winds. Meanwhile, crouched at ground level, baby sister Squash spreads her large leaves, providing cooling shade to prevent the surrounding soil from drying out as well as increasing humidity to help with bean pollination and suppressing competing weeds as she grows.

An added bonus, as veggie growers will know, is that beans can perform a nifty trick – taking nitrogen from the air and routing it back into the soil. It's a proven method of harnessing nature to do some of the heavy lifting for you, and besides, inventing your own holy trinity is a fun way of finding out which plants get along. Try growing lettuce, tomatoes and garden cress together. Taller bush tomatoes provide shade for lettuce and cress that like things cool and keep woolly aphids off your tomatoes. Marigolds tucked in next to greenhouse tomatoes work wonders in deterring green and whitefly, and simple hacks such as growing shallow rooters like onions alongside deep-rooting carrots (neither of which cast heavy shade) means less competition for light, while the onion scent throws inevitable carrot root fly off the trail of your main crop.

Chervil distracts slugs from more precious produce; dill patrols against hornwort – and suddenly, all's right with the world. Happily, mixed communities are less vulnerable to pests and outbreaks of disease, so you can be confident that diversity provides plants with greater resilience to nasties, if and when they do occur. Nasturtiums are a delightful, unruly mob that can't seem to work out whether they are climbers or scramblers. Tumbling in careless rapture, they lure flea beetles away from cucumbers and squash, and with gorgeous sunny faces and personalities to match, they prove effective decoys planted alongside cabbages, challenging any assault from cabbage butterflies and their voracious offspring.

Head over to any West Indian banana plantation and you'll find garlic hanging in trees to ward off damaging pests; a row of garlic or chives in a domestic garden is equally effective in dissuading blackfly and black spot from roses. And much has been touted about the benefits of planting chamomile alongside sickly individuals as a health tonic. Herby companions such as lavender, dill and fennel attract enough hoverflies to hoover up thousands of pesky aphids, while basil, coriander and parsley appeal hugely to ladybirds, who are more than happy to gobble up your aphid surplus, depleting the pest population by 5,000 per head in their lifetime. And anyone can easily make a brilliant DIY fertiliser by stewing water with comfrey or nettles in a barrel, both naturally high in nitrogen – it works like a charm every time.

Unsurprisingly, plants can fall in with the wrong crowd, too. There have been rare times in my garden that despite my best efforts – in this case, lettuce – and giving it the best five-star treatment available, my salad crops quite suddenly refused to grow. Having prided myself on being a champion salad grower, and after discounting the usual scourge of slugs and snails and exhausting all probables – more water, less water, less sun, more sunlight, more shade, pests and diet – I was forced to conclude the blame might lie elsewhere. After some deep-rooted browsing, one suspect kept resurfacing again and again. Allelopathy – chemical warfare raises its ugly head again – the ability of one plant to influence the growth of another for good or bad. Once you've witnessed it in real time, you'll begin to understand what a potent force it is.

I don't usually like to single out individuals for fear of besmirching reputations but in this case, I'm quite happy to make an exception. Goldenrod, or *Solidago canadensis* to the Latin-inclined. To be fair, I've never been a great fan. Its flowers are brash, more dirty mustard than gold, and despite its height, it's a slouchy, lolloping thing that is forever turning up where it's not wanted. My veggie patch was doing very nicely until it self-seeded near newly planted rows of pristine lettuce but once this scheming goldilocks turned up, things turned nasty.

We've been discussing happy bedfellows that aid pest control, enrich the soil and generally keep conditions favourable, but as it turns out, goldenrod is a malevolent thug. Individual stems can produce over 10,000 seeds, which accounts for its despotic inclinations, and as if that wasn't enough, its roots spread in a pincer movement underground while its height gives it a natural advantage, soaking up sunlight while depriving its next-door neighbours. The culprit in question also appears to have a high pollution tolerance, which explains why it thrives across so many different regions from cultivated gardens to scrub and industrial wasteland. While grudgingly acknowledging that even the most flawed among us have positive attributes – one has to admire its resilience – this does not extend to salad. The wretched thing wreaks chemical warfare that inhibits the growth of other species, including radish and *my* lettuce, which negatively affects germination. In hindsight, my poor old lettuce didn't stand a chance.

Happily, it's not just in domestic spaces we are seeing a return to companion planting. Modern agriculture has long favoured a monocultural tradition, growing single crops like wheat or potatoes, which facilitates large-scale farming heavily reliant on chemical fertilisers and pesticides. Hedgerows that in yesteryear provided windbreaks and wildlife sanctuaries have been ripped out to accommodate more cost-effective prairie farming.

It makes it easy to understand the devastation of the nineteenth-century Great Irish Famine, which saw thousands of families starve to death. But it effectively highlights the dangers of single cropping. Fortunately, greater environmental awareness of the benefits of shared resources, where different plant species are grown together, has prompted a growing trend towards polyculture and biodiversity in encouraging a healthier, more stable agricultural environment. Echoing the sentiments of one of my favourite authors, Alfred Austin, an eighteenth-century British gardener who couldn't have demonstrated a truer wisdom, 'Exclusiveness in a garden is a mistake as great as it is in society.'

CARING:
Plant Husbandry

With a better appreciation of how plants have evolved and their struggle to survive, a little plant husbandry can only enhance their well-being and yours. Plants are conscious beings. Every time you prune and snip, water or pick blooms for the vase, you are communicating with them. And they react. Brush against a nettle and, as we've seen, it triggers an immediate defence. Add organic matter to your flower borders and suddenly everything greens up and shoots within days. Mow the lawn and it quickly regrows. It stands to reason if you wish to grow anything with a fair degree of success, you must first understand it. Hit-and-miss planting will never elevate your garden from the ordinary to the sublime, but gaining a genuine cognizance of what plants need makes the real difference.

Growing isn't reality TV where a slick makeover is affected overnight. Like learning any new skill, gardening is an accomplishment gained over years, not weeks, so first off, you'll need to cultivate patience. Easier said than done for most of us, but nature refuses to be hurried. She's got it down pat after centuries of experience so it's best to grow a little fortitude alongside your crops and flowers.

I've always believed, misguidedly as it happens, that anyone who has been gardening for a few years would have accrued a wealth of knowledge. An assumption overturned by the fact that having given countless gardening talks, I've come to realise that gardeners roughly fall into three categories: those who stuff their trolleys full of bedding plants and toffees at a garden centre the moment there's a glimpse of spring sunshine; those who have specialised knowledge of particular plants; and gardeners who are a bit more thoughtful about the process. A veggie grower might know all about cabbages and onions but since they grow little else, they never really broaden their horizons. Newbies plant sun lovers in the shade, the ubiquitous fern in a sizzling hotspot, and all too commonly, you'll see clematis strung up and left to fry on a searing porch. Rather like children, plants can't choose their parents, so try to be, if not a knowledgeable carer, then a good enough one.

It's surprising how few gardeners take time to understand what is really going on in their gardens, rarely troubling themselves with plants that might be misbehaving or acting out of character. Plants are pretty vocal in their own way: some flower too soon, others not at all, some swoon, while others sulk. If they're unhappy, they show it. Rusting leaves, mean flowers, spots and pimples, foliage riddled with holes, felty fungus and powdery mildew are just some of the ways they express their anxieties.

The old adage 'right plant, right place' should be the foundation of every gardener's education. Within reason, if you replicate the conditions of a plant's birthplace you're more likely to enjoy success – and without a shred of doubt, become a better plantsman along the way. Where does the plant come from? Does it like heat, sun or shade? What sort of weather, soil or local hazards would it be exposed to in its homeland? Get a firm grip on the fundamentals and sensibly relate that to the kind of home you're offering and you can't go far wrong. Put sun lovers in the sun, shaders in the shadows, but try to force a sun worshipper to grow in gloom and you're asking for trouble.

Test the system if you like. Go ahead, I dare you. Plant lavender in a soggy flower bed in an inhospitable north-facing spot and at first they'll look just fine. Underground they've launched into survival mode, exerting all their energy into getting their roots down. It might be a glorious Indian summer, but suddenly there are downpours, their feet get soaked in foul water, and with so few hours of sunlight, they can't warm up fast enough to make energy, let alone dry out. Root rot and mould is their way of saying, 'I'm a celebrity – get me out of here!' And if rainfall doesn't kill off your silver-leafed sun worshippers, a chilly, wet autumn will. Try to see it from a plant's point of view. Mediterranean in origin, lavender, santolina and other silver surfers are accustomed to a spartan diet of sun-baked, stony soil and minimal summer rainfall, so don't spoil them. It's a rich diet for plants used to being malnourished and whilst you might not be exactly mimicking the romantic lavender fields of Provence, you should be gunning for conditions as close as possible.

Realising the importance of soil in creating a potent infrastructure that holds even the smallest cultivated plot together, it pays dividends to look after your garden environment. Your plants are only as good as the soil in which they are planted – and don't let anyone tell you otherwise. It's astonishing that people will buy a shedload of healthy plants then immediately plunge them into austerity. How would you like it if every time you opened the fridge there was nothing to eat but a mouldy jar of jam and a carton of sour milk?

A case in point is illustrated by a lovely friend who asked my advice on planting an ornamental hop. If you've ever grown a golden perennial hop, you'll know that once it has decided its new home is satisfactory, it gallops away a mile a minute. 'Give it a nice sunny spot, decent soil and regular watering,' I advised. Three weeks later, she complained it was as dead as a doornail – an unspoken accusation between us, as if I was responsible for its untimely demise. It turns out she'd decanted the unfortunate plant from its plastic pot, filled a planter with fresh compost and set the entire thing on the surface and then, for some unfathomable reason, decided it could do without water. Now, there's a plant that regretted ever leaving the luxury of the garden centre.

Remember, plants growing together may seem complementary to the human eye, but the chances are you're growing at least a dozen single species in your garden, likely most of them unrelated. It's pretty decent of them to rub along with rival tribes for our pleasure and nine times out of ten you can get away with growing incompatible sorts as long as you cater for their individual needs.

Trying to keep a disgruntled plant happy in soil that contradicts its natural requirements is a thankless task. If you don't have acid soil, give acid lovers such as azaleas, blueberries, camellias, heather and rhododendrons a miss. Acid lovers are prickly curmudgeons if grown in incompatible soil. Foliage turns a sickly yellow as they grudgingly squeeze out meagre flowers, a blatant reprimand that they prefer their toes in a more compatible blend. With thousands of alluring plants on offer, you're not missing out, and can easily find worthy alternatives. Gardening should be fun, and creating a happy outdoor environment not only benefits your plants but goes a long way into making us happier people, too.

HAPPY SPACES:
Flower Failure, Latin and Trouble-Shooting

A common way plants express their anxieties is through telltale yellow leaves – a sure sign that a plant is ailing. It's usually flagging up a nitrogen deficiency and, as we know, plants make energy through their leaves, so sickly foliage seriously impairs that ability. Never mind. Nitrogen deficiency is one of the easiest ailments to remedy and plants usually green up pretty swiftly with a nitrogen boost. That applies to lawns, too. Look for the letter 'N' on fertilisers. Pelleted chicken manure, mushroom and home-made compost or good old-fashioned, well-rotted farmyard manure keeps outdoor larders well stocked with vital nutrients throughout the year, negating the need for additional fertilisers. Not to mention it saves you time and money helping retain soil moisture, so it cuts down the time spent on watering regimes, too.

We've talked a great deal about plant structure, and in studying your garden charges you'll quickly understand quite a bit about what any plant needs to be happy just by looking at it. Notice how leaves are arranged around stems, or the size and texture of the foliage. Large leaves often indicate a preference or at least some tolerance for shade, given larger leaf areas can absorb greater amounts of sunlight. Do the flowers bloom before or after the leaves? And even flower colour will tell you pretty accurately what a plant needs to thrive. Grasses rely on wind to sow their wild seed. Bright-coloured blooms have their hearts set on attracting bees, hummingbirds, hoverflies and butterflies, whereas those that begin to waft scent as dusk falls tend to have pale blooms, exuding a seductive twilight luminosity intent on wooing night-flying moths or bats.

Strange as it may seem, plants can resist flowering, and it's normally down to two factors: natural causes and human error. Low winter temperatures and frosts nip plants in the bud, destroying any potential blooms, and unless a plant can throw up new flower buds pretty sharpish, they can go an entire season without flowering. Disappointing, but hardly fatal, so don't worry. Besides, nature throws us all the second chances anyone could hope for. In response to long nights (or short days), lack of sunlight may cause flower-blindness, where foliage looks hale and hearty but flowers are never formed down to sun scarcity. Sometimes old age accounts for flower failure and typically shade reduces the abundance of flowers in many sun-loving species. Many plants that aren't natural sun worshippers are sold by garden centres as happy flowering in shade, but trust me, their blooms are greatly reduced in the shadows. Having said that, I guess we all have to make comprises, and it's better to have a few beautiful flowers than none at all. However, if you have a deeply shaded or north-facing garden, go for gloom lovers that naturally bloom in shade.

And we shouldn't overlook the cack-handed gardener who is often an unconscious instigator of flower failure, pruning buds too early, being stingy with the watering can, planting things in the wrong place at the wrong time or expecting plants to do too much, too soon. Many woody plants need to reach maturity before putting out flowers or fruit. A gingko tree can take fifteen years before it's ready to bloom, while trees such as apples and pears need at least three or four years to put out any meaningful fruit, as do apricots and peaches. Understandably, young fruit trees need to develop a strong framework of branches to hold the weight of a decent crop.

If you're new to gardening, don't expect to know everything in a nano-second, you'll work things out as you go along. Look at it this way – with thousands of plants on the planet, even if you lived hundreds of years, you'd never have the time to grow them all. As in life, you'll meet some fascinating characters; some will enthral, others are simply horrifying, and although not everyone is going to be your cup of tea, I guarantee you'll have some interesting experiences along the way.

Even plants growing in the *right* place in the *right* soil suffer burnout through depleted food supplies, adverse weather, water shortages or benign neglect by their owners. A woeful list of calamities triumphed by an innovative Canadian gardener who asked my advice on how to protect her vulnerable plants through savage winter snowfalls. Fleece, I advised, wrap them in warm fleece. An email arriving in my inbox a couple of weeks later with attached images demonstrated her dedication. Jauntily insulated against winter cold, all were wearing hoodies! (I hadn't thought to emphasise horticultural fleece.) You'll be pleased to know her plants came through just fine, and I'll bet they're the most stylish plants on the block.

I guess my point is that you are better taking advice from experienced gardeners or nurserymen than commercial companies that have a vested interest in profit over real-life scenarios. I'm reminded of a stroll through the Amsterdam flower market, where an impressive 6ft pillar of plump, ripe strawberries glistened in a tantalising tower of deliciousness. My sister was about to hand over her euros for what the Dutch proprietor swore defiantly was a climbing variety when I stayed her hand. 'Save your money,' I laughed. 'There's no such thing as a climbing strawberry.' Naturally, strawberry runners can be trained to climb short distances with the right scaffolding, however, I've heard all kinds of inducements in my time from plant sellers who would have us believe any kind of nonsense when it comes to making a sale. If it looks like a duck, sounds like a duck ... then it probably is a duck.

'The difference between the almost right word and the right word is the difference between the lightning bug and the lightning.' Mark Twain might as well have been taking about plants. When I started out in horticulture thirty years ago, I'll admit plants were baffling. What I knew about them, their origins or requirements would have fitted neatly on the back of a postage stamp with plenty of room left over. There's no shame in not knowing the difference between a geranium or pelargonium, so as we progress on our botanical adventure, let's get some of those bewildering Latin names out of the way.

For those unfamiliar with nomenclature, like us, plants belong to families, and those often baffling labels are simply a way of describing which clan they belong to. In the eighteenth century, a Swedish biologist, Carl Linnaeus, decided it would be a good idea to give unique species a double name comprising the genus (the family name, like Smith or Baker) and a species name, which provides the more descriptive aspects of each family member. *Uncinia rubra* (red sedge) simply translates as belonging to the sedge family and its leaves are red. Or in layman's terms, Joe Baker – carrot-top or redhead. Similarly, winter honeysuckle (*Lonicera fragrantissima*), a gorgeous, arching, winter-flowering shrub, belongs to the honeysuckle clan and is perfumed. *Armeria* is the genus name, *maritima* tells us the species so you know plants with this description enjoy seaside locations, and 'Alba' simply tells us this dainty, tough thrift has white flowers.

Latin names are incredibly descriptive, so it's worth checking out those plants labels, since they all give valuable clues to a plant's colour, preferred location, size and shape. I'll leave you to guess at these examples for yourself: *arborea*, *aurea*, *compacta*, *contorta*, *foetida*, *gigantea*, *horizontalis*, *pendula* and *sanguineum*. How did you get on?

It was an innovative and staggeringly ambitious concept, since plants are known by different nicknames in different countries. With Linnaeus' system, finally, we could all be sure we were talking about the same plant by using its Latin epithet in whatever country we happened to be. You can imagine nomenclature is also incredibly useful considering the medicinal properties some plants contain because even if they belong to the same family, some may be more toxic than others. Frankly, it's all too easy to court disappointment for lack of botanical correctness, and if you were hoping for a fragrant drift of rich purple lavender and instead happen to pick up two dozen random lavender at the market that bloom to wishy-washy pink, it completely decimates any vision of a romantic violet haze.

All apprentices make blunders while learning their trade, and floral howlers are par for the course but as your interest in nature develops, you'll begin to realise that you can confidently tackle your plants' anxieties while alleviating your own.

NURTURED BY NATURE:
Stress Less, Live Better

In 1859, Charles Dickens set about writing his empirical novel, *A Tale of Two Cities*. I mention it simply because the sentiment expressed in the opening chapter is prophetic – as relevant today as it was back then:

> It was the best of times, it was the worst of times, it was the age of wisdom, it was the age of foolishness, it was the epoch of belief, it was the epoch of incredulity, it was the season of Light, it was the season of Darkness, it was the spring of hope, it was the winter of despair, we had everything before us, we had nothing before us, we were all going direct to Heaven, we were all going direct the other way.

The path to happiness seems to be the holy grail of the modern age. That intangible something hovering around the next corner and always tantalisingly out of reach. Lockdown magnified that fragility profoundly.

It was fine at first, wasn't it? COVID triggered an optimistic frenzy of homemaking. During the opening weeks we embraced home working and new hobbies as a bit of a novelty, labouring under the misapprehension that all we needed was to learn a few new skills to see us through what we imagined would be a temporary confinement. Little realising it would go on for months, across the world people rediscovered traditional pursuits from baking, cooking, candle-making, bread and jam-making to writing and gardening. With offices, gyms, leisure centres, restaurants, bars and pubs, holidays and socialising off limits, the distractive luxury of pursuing interests previously deferred or overlooked seemed something of a blessing. However, living, working and home-schooling our children piled on unbearable daily pressures for millions. Young and old, rich and poor – the world suffered largely in isolation. Never in peacetime have so many had their physical liberties so drastically curtailed. Unsurprisingly, enforced captivity triggered a natural human instinct for rebellion, escape and ultimately freedom.

You've heard of the phrase 'touch grass' – another way of saying 'calm down' – and nature offers its own breakaway; periods of calm and physical movement where for a few precious hours we can set aside life's troubles. After nearly four decades of gardening, I'm convinced that nature has the power to teach, to inspire and to heal. Not always offering magic-bullet solutions but enough to jog us on from those dispiriting times and get us, largely intact, however flawed, to the next junction.

Recent research supports this. A survey by Censuswide found that 63 per cent of UK adults reported improved mental health through gardening. The Royal Horticultural Society reported similar outcomes: 80 per cent of people said gardening improved their overall fitness, 60 per cent felt more energised and 53 per cent reported an uplift in mood. That's a remarkable return for such an enjoyable activity.

Health practitioners worldwide are increasingly recognising gardening as an accessible form of therapy, one that boosts life satisfaction, enhances energy and encourages positive thinking. Studies show long-lasting benefits for those with depression, anxiety or ADHD, while horticultural therapy – a rather clinical phrase for a life-enhancing practice – is used successfully to support military veterans suffering from PTSD. Gardening gives them focus, calm and a way to work through their inner battles, and if it can help those facing such profound challenges, imagine what it can do for you. In the course of writing this book, I've met people from all walks of life who struggle with physical or mental health difficulties. Again and again, gardening has proved to be a lifeline: accessible, grounding and deeply restorative. Thrive, a UK-based charity, uses gardening to bring positive change to disadvantaged, isolated and vulnerable people, as well as those suffering from dementia and poor physical and mental health. Here are just some of their experiences:

> When I spend some time in my garden, I'm stepping out of the things that have gone on, I'm not in the past.
>
> –Abi, on overcoming PTSD and anxiety.

> When I finish a session, I feel good; I've strengthened my body and I've strengthened my mind and my motivation. I feel a part of something great and so important. I get great satisfaction and I'm keen to learn new things about gardening.
>
> –JoJo, after being discharged from long-term psychiatric care.

> At the end of the session, I feel happy, invigorated and that I've learned something new.
>
> –Cynthia, with early dementia.

It took a pandemic to make us realise that perhaps the way we ran our lives was far from satisfactory. Isolated, socially deprived, emotionally bruised and vulnerable, we drifted on unfathomable seas with no happy ending in sight. Cut off from the very people we could normally rely on to keep us sane, is it any wonder we increasingly yearned for greater freedom, an escape hatch from mind-numbing incarceration? Property prices hit a premium as homeowners competed fiercely for more desirable residences with outdoor space as we traded the confines of urban living and flocked to the country or headed for the coast in search of a better quality of life. Up until then, few of us really appreciated nature's potency, much less appreciated its potential for lifting us out of depression, restoring positivity and bringing hope. Subconsciously, we were drawn outdoors; nature became our lifeline to sanity.

Increasingly, medical experts endorse interacting with nature to boost our physical and mental health. Natural spaces reduce stress, promote positive mood and would you believe that hospital patients with views overlooking natural settings healed faster than those without? It's no coincidence that millions of Americans, Asians and Europeans flock to parks and beaches to join various physical activities in the open air – events that are invaluable to seniors, encouraging them to maintain regular fitness and social interaction.

Unfortunately, it's human nature to slide back into old habits, and recent figures show that at the beginning of April 2024 over two-thirds of the world's population (5.44 billion) are connected to the internet for either work or leisure. The hours we spend on technology are staggering. The average US adult clocks up ten hours a day. British adults averaged over six hours daily, and more worryingly, our kids spend weekly totals of eleven hours gaming, over fourteen hours on their mobile phones and thirteen hours watching TV. Naturally, the plus side is that people across the world are connected in a way that was inconceivable before the advent of the worldwide web that was publicly launched in August 1991. But it's the amount of time we spend that's the niggling issue.

You don't need me to explain that your immediate environment can either increase or reduce stress or can directly affect our mental and physical health. An unhealthy home, workplace, lifestyle or relationship makes us unhappy, leaving us feeling depressed and helpless – a sense of entrapment with no obvious exit. In contrast, a natural environment promotes contentment, security and gratitude. Research shows that nearly two-thirds of us seek out nature when we need to retreat from our busy lives, whether it's cycling, jogging or simply taking the dog for a leisurely walk.

The great outdoors is where we turn for relief from the pressures of work and home. Modern stress inevitably elevates heart rate, blood pressure and muscular tension. Over time, it builds up, suppressing our immune systems and leaving our minds tangled and tired. Nature has the opposite effect. It soothes and reduces stress hormones, calms heart and blood pressure and replaces negative emotions with positive ones. Without question, spending time among nature makes you happier. Birdsong at dawn, the smell of freshly cut grass, warm earth crumbling through your fingers, getting drenched on a wet, windy spring morning – moments that connect us to something priceless. Small pleasures with immeasurable health benefits. The American botanist Luther Burbank summed it up nicely, 'Flowers always make people better, happier, and more helpful; they are sunshine, food and medicine for the soul.' In other words, every moment in nature is a deposit in the health bank and it can make us better parents, too.

PARENTING:
Backed by Nature

Was it Margaret Atwood who extolled, 'In spring at the end of the day you should smell like dirt?' Who am I to disagree?

Dirt and fresh air are a heady concoction, and as we grow into adulthood we seem to have less and less to do with both. As kids, dirt was obligatory. It was under our fingernails, grubbed our knees, and in my humble opinion, even in adulthood most pursuits are improved with its inclusion. If you happen to stumble across me in outdoor dishevelment, mud-stained jeans, sweater bobbled with sticky seeds and hands streaked with dirt, the chances are you'll find me in the best of moods. Everyone should spend a few hours a week grubbing around in the dirt. It's liberating and frankly, if I were in charge I'd make it mandatory.

I recall my daughter as a toddler, her face screwed in dismay when her ice cream slid off the cone and hit the pavement – splat! To the disgust of a horrified mother standing nearby, she hit the deck and lapped up every last scrap. While not advocating pavement dining (although, to be honest, I've been in less-promising eateries), ongoing research into childhood illnesses argues that dirt deficiency might be making our kids more susceptible to irritations, asthma and allergies.

In what was already a hygiene-obsessed society, COVID did little to diminish our germ paranoia, but it appears there are few things more effective in delivering a jolt of joy and immunity than a handful of dirt. Getting your hands caked in mud does wonders for your happiness quotas because a soil-based bacteria, *Mycobacterium vaccae*, triggers dopamine, the happy hormone! So, it rather begs the question why more of us aren't spending a bit more time in the brown stuff? If we are to sensibly achieve natural harmony, we need to start from the ground up, and what better way to begin than by focusing on our children?

The joy of play is much underestimated. I'm talking about the imaginative, no-rules, do-what-you-like kind of play that gets left behind in childhood. Gardens and natural outdoor spaces promise an exciting environment that validates dirt collecting, cloud gazing, chasing squirrels and climbing trees. Piles of rusty autumn leaves present the perfect opportunity for leaf angels every bit as much as fallen snow, or how about splashing in puddles?

Concerns about nature deficit posed increasing challenges long before we were held hostage by a pandemic, and a lack of physical freedom has a knock-on effect. Almost every country worldwide has seen a steady rise in obesity across all age groups. According to the World Health Organisation, obesity in kids under 5 living in high-income countries has seen an increase and, worryingly, the trend is expected to accelerate. In England alone, one in three kids leave primary school overweight or obese, and in North America, 25 per cent of pre-schoolers are already above their ideal weight.

Public health experts confirm that poor access to outdoor spaces is a key contributor. If we continue to let our children grow up overweight, you don't have to be a rocket scientist to realise it puts them at a higher risk of developing mental and physical health issues. Allowing our kids the luxury to explore the great outdoors and muck about in the dirt through simple play is an evidence-backed healthy regime.

Connecting your kids to healthy soil is a ridiculously easy thing to do and offers an impressive roll call of positives. Making mud pies and chasing about in the greenery has an anti-inflammatory effect on young brains and boosts academic achievement, as well as self-confidence and self-reliance. A Finnish study on city-based children attending a daycare where a living forest floor had been installed found they had stronger immune systems than those kids faced with hard outdoor-play spaces. They also had healthier microbiomes and continued to show beneficial gut and skin bacteria some two years later. Shunning sterile, concrete playgrounds of more traditional schools, long-term subscribers to the benefits of *Naturebørnehaves*, or forest schools, which were first conceived in Denmark in the early 1950s, encourage kids to play outdoors for around four hours a day, come rain or shine. Scandinavian kids thrive on imaginative play, fashioning tools from branches and twigs, playing hide-and-seek and building shelters, giving them license to explore the natural world for themselves. With an emphasis on adult assistance rather than supervision, children park indoor classrooms and technology to roam woodlands, climb trees, catch bugs and light camp fires. You might imagine that a dozen children left to roam in woodlands would be pandemonium, but as it turns out, noise levels at forest schools are lower than most conventional ones. With an emphasis on social, emotional and technical development through learner-initiated learning, forest-schooled kids benefit steadily from increased memory recall and attention spans, as well as improved cognitive and problem-solving skills. Nature play heightens curiosity and sociability and, just as importantly, helps our kids calculate risk. Do you mind if we talk about risk for a moment?

In case you were wondering, plants assess risk, too. Every second, plants make critical judgements between defence or investing in growth. From the moment a seed sprouts and the first leaves emerge, any number of events threaten potential harm. They can be eaten, flattened, harvested or an insect bite could prove deadly. Aphids, caterpillars, slugs and snails, animals, potential disease and adverse weather – the threats are legion. Naturally, once a plant has reached a reasonable size it can afford to lose a few leaves to munching critters because the remaining leaves can still make sufficient energy to compensate for any damage but twenty-four-hour vigilance helps mitigate the risks.

Michael Douglas' character, Gordon Gekko, in *Wall Street* famously extolled, 'Greed is good' and I'd argue the same for reasonable risk. Cast your mind back to childhood – daring your friends to climb the highest tree or toboggan down the steepest slope (usually on a tea tray nicked from the kitchen), we experienced real fear followed by pure exhilaration.

From a layman's point of view, I'm assuming risk is largely driven by two prime factors: wanting to do something positive in our lives with the awareness it may fail or is potentially injurious, or by avoiding taking any risks at all, neatly safeguarding ourselves against failure, injury or disappointment. I'm sure psychologists could distil this simple argument into a much broader conversation, and without denigrating current health and safety guidelines, which have certainly improved working conditions and other areas of our lives, some regulations increasingly curtail our freedoms and blunt our natural instincts. Surely taking chances can be a good thing?

We're taught to reach for the stars, to strive for better, yet how can we do that hobbled by risk avoidance? There isn't a best-selling actor, author or sportsperson who put themselves out there out there without the dread of rejection. Antarctica wasn't discovered because courageous explorers decided they might be safer at home warming their slippers by the fire. Gut instinct, like every other part of the human body, requires constant exercise to maintain it and while I'm purposely overexaggerating, gut instinct is there for a reason and most of us are subconsciously aware of where dangers lurk. And how did we get there? By steadily building our risk sensors through childhood.

Nowadays, children the world over play outdoors less than their predecessors. Understandably, parents fear 'stranger danger' and let's face it, indoor activities like watching TV or playing computer games are more easily supervised and keep the kids clean and tidy. There's no denying indoor activities stimulate kids' minds in other ways, but exploring natural outdoor environments in early childhood is vital for physical, emotional and mental development. Nature is a wonderful, free resource, allowing our children to journey into their imaginations and creativity – a fascinating place where they can burn off energy and test their resilience by escaping into a world of adventure while strengthening young muscles and improving balance and coordination.

Even at my own prep school, Guy Fawkes Night was an eagerly anticipated annual event. Each child was handed a foil-wrapped raw potato and a stout, straight branch, its tip bound in cloth dipped in paraffin, while being firmly warned by our teachers that fire was dangerous – *Don't you go waving those about torches, now! Stay away from the bonfire!* We happily obliged for what we considered a worthwhile pay off – the daring of being out past our usual bedtime. Holding flaming torches aloft, fireworks whooshing overhead in a November sky, we marched excitedly to the bonfire. As the blaze died down, the highlight of the evening climaxed in cooking our potatoes in its warm ashes. To the best of my recollection, none of us ever sustained injury or burns. Today, this kind of activity would be deemed far too risky for young children, so I consider myself lucky to have enjoyed a childhood where experiment and experience were just as valid as being able to recite multiplication tables or rattle off the kings and queens of England in chronological order.

As parents, we always try to put our kids' interests first, but how about a little me time? A proven fitness regime that costs nothing and is guaranteed to make you healthier and happier.

GREEN FITNESS: Natural Workouts

Unsurprisingly, I have an unashamed bias when it comes to the benefits of gardening, which effortlessly combines relaxation and physical exertion with mindfulness, quite capable of chasing away the problems of the day. I'd bet good money on it that you're less likely to abandon a relaxing hour or two in the garden than the tedium of treadmills, bench presses and clock watching. (Stomach crunches always induce fevered mental calculations in my case.) What's more, health experts agree: gardening is an affordable, accessible activity that boosts fitness and life satisfaction. It makes you fitter, speeds weight loss, promotes better sleep and reduces anxiety.

It's not just my opinion – the benefits are backed by medical evidence. Dr Bird, adviser to Public Health England, puts it simply:

> Gardening is a physical activity which can help reduce the risk of obesity, type two diabetes and heart disease. It's a great exercise for burning calories and for getting people outdoors and close to nature, which makes people feel psychologically energised, too.

There you have it. Regular contact with nature alleviates depression and anxiety, improves circulation and favours longevity. Best of all, it's non-exclusive, embracing young and old, those suffering with disability or depression, transcending age, gender, race, culture and religion. Interestingly, UK health professionals are urging the NHS to prescribe gardening more often, noting that every £1 spent on outdoor community schemes could save £5 in alternative treatments. That's a return worth getting the spade out for.

According to a UK government poll, because of the cost-of-living crisis, 10 per cent of British adults cancelled or are considering cancelling their gym memberships. Added to that, some 67 per cent of gym memberships (no doubt initiated under a swell of well-intentioned New Year resolutions) go unused. Reasons vary: people are just too busy to make time to attend; queues to access equipment are off-putting; a lack of will power or self-confidence; or just the whole business is a bit of a faff. Fortunately, connecting with nature isn't difficult. Gardening time is happy time, offering positive results where you stress less and smile more – a solid workout, medically proven to build muscle and burn calories but with extra benefits. Just one hour spent outdoors lowers stress and lifts mood. And we shouldn't underestimate the positives of having a chat over the fence with a neighbour or standing hoe-to-hoe with fellow allotmenteers to give a sense of purpose and belonging.

Can gardening really keep you fit? Hell, yes, but don't just take my word for it. Committing to just thirty minutes of moderate gardening three times a week promotes better sleep, weight loss and overall well-being – without expensive equipment or fees. And happily, it's easily timetabled into our leisure hours. Low-intensity tasks such as carrying, pruning or weeding increase the heart rate. A resounding 70 per cent of gardeners claimed it toned forearms, 52 per cent said it toned thighs, while a third said it toned bums. All that bending, crouching and stretching keeps us supple, helping to maintain overall general fitness. Based on figures from Harvard Health Medical School, for an average person weighing 11.7 stone (155lb), moderate activities such as mowing, digging and planting slims waistlines, smooths love handles and tightens buttocks.

So, what's your preference? Digging your plot or a step machine? Both burn around 200–400 calories hourly. Hatha yoga burns 144 calories per thirty minutes, while raking or weeding expends the same number of calories, depending on how long and how much effort you lend to the task. And weeding feels a whole lot less tedious when you realise it burns up to 200 calories in thirty minutes – and since weeds are relentless in their aspirations, it soon speeds up the fat burn. Plain old raking, bagging leaves or shovelling snow can burn up to 400 calories hourly versus low-impact aerobics, which burns only 50 calories more. And if you trade in your power tools for old-school methods such as push mowers and hand shears, you can ramp up the intensity.

Beyond the calorie burn, gardening relaxes the body and mind. Under grey or sunny skies, pulling potatoes or picking cherries guarantees you'll feel brighter and lighter. Whether sowing seeds or just turning the compost heap, you'll find it's the most relaxing lifestyle imaginable – good for body, mind and soul. Yoga for the mind, if you like – a natural pick-me-up that builds confidence, creativity and fosters a quiet sense of achievement. And unlike people, plants never criticise your efforts. Their faith in you is absolute, even if it is occasionally misplaced.

If you're still sceptical, why not give it a go right now? Switch off your phone, turn off the TV and step outside. Or perhaps you fancy a walk instead?

WALKING:
A Wonder Drug

Hippocrates was one of the first to extol the virtues of walking, 'Walking is man's best medicine', and increasing awareness among international health practitioners demonstrates a resounding endorsement of his dogma. According to the leading medics, walking is the nearest thing we have to a wonder drug.

It's all too easy to dismiss the benefits of something as simple as a stroll, but next time you visit the doctor, wouldn't you be pleasantly surprised if they prescribed taking a walk instead of medication? The fabulous thing about walking is it's free and almost anyone can do it. Ralph Waldo Emerson lists the necessities of a good walk, 'Few people know how to take a walk. The qualifications are endurance, plain clothes, old shoes, an eye for nature, good humour, vast curiosity, good speech, good silence, and nothing too much.'

Walking, the only permitted form of exercise, became an international pastime during lockdown; even if we couldn't touch each other, at least we got out in the fresh air to rub shadows if not shoulders with others in that alien landscape. In a bid to escape confinement, walking became our lifeline to sanity – a much-underrated form of exercise, offering a whole host of benefits, treating mind and body in one fell swoop.

With a positive effect on both emotional and mental health, walking out in the fresh air triggers the release of brain-derived neurotrophic factor (BDNF), a special protein secreted by the brain that helps keep neurons healthy. I'll admit I'd never heard of BDNF until recently. Among its many functions, this miraculous protein untangles parts of our brain that have become broken or knotty, which is why walking instantly helps alleviate depression and guarantees an instant uplift. Instead of grabbing a cup of coffee, a quick stroll is likely to dissipate anxiety more effectively than a caffeine fix. Ditto fatigue, stress and anger. The moment you make a commitment to walking on a regular basis, it stands to reason you're going to feel a whole lot better, more of the time. My sister is an enthusiastic advocate, always ready with a well-worn cry, 'I've never met a walk I didn't like!'

Walking can be solitary or sociable, and a brisk circuit around the block during your lunch break does wonders for creative output, too. Human creativity is heightened when we're moving as opposed to being sedentary. It's claimed that gardeners and florists are the happiest professionals, with nine out of ten claiming they are happy in their jobs. Bankers scored lowest on the happiness index with only 44 per cent content with their lot. Unsurprisingly, emergency workers and first responders are often the most challenged in trying to establish a regular fitness routine because of the uncertainty of daily schedules, so for them, perhaps an impromptu walk is more easily doable. A growing band of woodland therapists encourage their clients to walk and talk, believing harmonious woodland encourages the mind to relax and help us speak more openly. Military personnel accustomed to being physically active or suffering with PTSD have found walking a lifeline to recovery, allowing them to re-engage with exercise, and a valuable tool to reduce stress, anxiety and negative thinking.

Walk the dog. Walk to lift your mood. Walk to work or to let off steam, or walk for absolutely no good reason at all. Just twenty minutes a day will significantly boost your immune system. This is evidenced by a study on committed strollers who walked for twenty minutes a day, five times a week and saw a whopping 43 per cent fewer sick days than those who exercised once weekly or less – and symptoms were milder when they did fall ill.

The American Cancer Society gives walking a mighty thumbs up as playing a key role in reducing breast cancer. Results showed that women who walked more than seven hours weekly had a 14 per cent lower risk of breast cancer than those who only managed three hours or less.

Walking provides a gentle, effective, sustainable drill for seniors or those worried about joint stress by strengthening cartilage. And the thicker our cartilage, the better it supports our joints. Humans are 100 per cent dependent on mitochondria, battery cells that power our everyday movement, which are designed to be in constant motion. When we remain static, it stops our batteries working effectively, so even short, regular spurts throughout the day helps boost efficiency.

I've discovered that walking is brilliant for writer's block. Over the years, I've fine tuned a double-quick routine guaranteed to unravel any obstacles and make you feel instantly upbeat. Just a two-minute brisk walk followed by a two-minute sprint, every two hours. Try it sometime. It stretches you away from hunching over the keyboard, is easily doable, burns fat, builds muscle and improves cardiovascular fitness as well as being time efficient.

Serial dieters acquainted with the misery of yo-yoing weight gain will be thrilled to learn that walking is one of the most-effective antidotes in any slimming routine. It counteracts the effects of weight-promoting genes by as much as half – burning five calories a minute. And chocoholics might find that willpower could prove a thing of the past because a fifteen-minute walk is proven to curb sugar cravings.

Yes! Good old-fashioned walking trounces statistics in every way, so it's hardly surprising this pedestrian lark has caught on. Call it a personal rewilding, if you like, increasing the likelihood of positive emotions, endowing us with energy, joy and strength.

Literature and walking are sworn bedfellows – there's hardly a poet out there who hasn't been inspired to pen a verse on the subject. From Wordsworth and Hardy to Housman, they have all sharpened the literary quill to wax on the joys of pedestrianism. Charles Dickens was said to be a rampant walker, religiously tramping from his London house to his country home in Kent, some 30 miles away, undeterred by distance wind, rain or fog. 'I think I must be the descendant, at no great distance, of some irreclaimable tramp,' he wrote, and considering the miles he clocked up it's a wonder he found the time to write at all.

My vote goes to Thomas Moore, the nineteenth-century ex-monk and spiritual writer, who's summing up in his book *Soul Mates* best describes an experience most like my own, 'Walking inspires and promotes conversation that is grounded in the body, and so it gives the soul a place where it can thrive.'

We haven't really spoken about the effect that olfactory stimulus has on human senses, yet floral fragrance plays a powerful role in our subconscious, softening worry into calm, making us happier, more hopeful human beings.

THE ESSENCE OF FLOWERS: Fragrance and Folklore

From Mother's Day bouquets to red roses on Valentine's Day, flowers are proven to trigger positive emotions. For a single species, roses possess a froth of fragrant notes. Notes of blackberry jam and citrus of heritage roses; the powdery spice of tea roses and the pungent cinnamon-clove musks, concentrating their fragrance in the stamens rather than the petals. Floral perfume is a mood enhancer, providing pleasurable bouquets that pique our emotions and as importantly, influence our dietary and cosmetic choices.

In the modern day, we spend a fortune on botanical boosters to perfume our homes or to enhance personal health and hygiene. According to the American Botanical Council, in 2024 consumers spent $533 million more on herbal supplements than the previous year, and the global botanical skincare market, valued at $1.22 billion dollars in 2023, is projected to double by 2031. Beauty products increasingly rely on botanical essences – oils of geranium, lavender and chamomile, basil and dandelion, essences of black pepper and white tea. While it's doubtful they have any convincing effect on wrinkles, saggy jowls or crepey necks, few would argue the blissful euphoria after a relaxing facial or body massage with flower-infused preparations.

Among the most expensive plant extracts is Oud, from a resinous tree in Asia. Highly prized for thousands of years, a tree needs to be affected by a particular mould to produce its unique, intense fragrance. Orris oil from purple-flowered iris comes second in monetary value, only because it can take up to three years to dry, powder and distil its essence. Jasmine trails in third place, on account that it can take approximately 8,000 flowers to produce 1/25 of an ounce of oil, but that is easily trounced in value by ambergris. Derived from the intestines of sperm whales, perfumers pay top dollar to extract an odourless alcohol from it called ambrein, which extends the scent of perfume.

Fragrance has practical applications, too. According to Nicholas Culpeper, wild Sweet William was a probable cure for pox. It's all too easy to sympathise with those snooty London physicians who thought him a charlatan, although at the time they had no better ideas for curing the plague. But given that sanitisation was poor and the majority of the world's population were peasantry involved in manual labour, fragrant plants played their part. Rose and thyme perfumed bathing water in the Middle Ages, long before bathing became essential toilette for society's elite. It's likely peasants made do by bathing in rivers during the warmer months or boiling pots of water over open fires in winter, while the well-to-do shunned frequent bathing, believing water was harmful to health by opening pores and leaving the body susceptible to disease. Despite having numerous bathrooms and running water, it's believed that Louis XIV only bathed three times in his life and history has poor old Marie Antoinette down as none too fragrant. Although unusually, she bathed more frequently than most of her era, adding prescriptive sachets of mallow, linseed and almond to her bath. Sadly, history books are coy about the frequency – shame, as a history buff, I rather enjoy that kind of detail.

Cutlery being the preserve of the wealthy, food was commonly eaten by hand, but it's reasonable to assume most people washed their hands and faces every day from a basin of water, trough or stand pump. Collected rainwater in rural areas would have provided a free source of water, though less so in towns and cities where river water was likely to be polluted. But even back then, people must have appreciated the benefits of overnight showers after a dry spell, awakening to a morning alive with that complex earthy tang. A draught so indescribably energising I wonder nobody has thought to bottle it. The ancient Greeks had a word for it – *petrichor* – that marvellous scent after rain.

At a time when straw bedding prevailed, fleas and lice were an inevitable scourge, and incorporating aromatic herbs such as chamomile, mint and lavender into the straw helped minimise the nuisance. Elizabethan theatres, welcoming the poor, middle classes and rich alike, were engulfed in a swampy miasma of unpleasant pongs. Standing areas, largely occupied by peasants, whiffed so badly it was enough to force the gentler classes in the pricier seats to adopt fragrant nosegays to mask the fug.

Contrary to popular belief, daily bathing wasn't the norm in Victorian times either, although naturally washstands and pitchers were employed on a regular basis. Bay, lemon balm, dried rose petals and sage were made into soap balls or fashioned into floral pouches that were slipped into clothing or held at the waist to mask nasty niffs.

With the introduction of the government's Baths and Washhouses Act of 1846, communal baths established by local authorities finally made regular bathing accessible to the working classes. Social history is always illuminating, and a frank tariff sheet shows what the average punter seeking to deodorise themselves was charged at a Glasgow public wash house that boasted it was able accommodate 200 washers:

> The Tacksman provides hot and cold water, 100 large and 80 small Tubs and 100 Stools to any persons who choose to apply for them, in consideration of the following: Dues, viz: Hot and Cold Water for a Day's Washing for one person, without the use of Tubs and Stools, Fourpence. Do for One Half-Day's Washing for one person without the use of Tubs and Stools, Threepence. One Day's Washing Tub, One Penny. Boiling clothes in large Boiler (one hour) Eightpence, do small Boiler (one hour) Fivepence.

Folklore holds a rich floral tradition, unique in its storytelling and passed down through generations, connecting us with plants in a sensory way that goes beyond their physiology. Long-entrenched in myth and magic, we find delightful examples of their significance through the ages. Latin names are all very well, but they sound a bit alienating, whereas the common vernacular of plants and herbs – bachelor's buttons, kitten-tail, lady's mantle and cuckoo spit – are so much more relatable. Old wives' tales going as far back as the early Egyptians show prized lotus flowers used as both an aphrodisiac and antidote to impotence. Boiled milkwort could increase a new mother's flow of breastmilk, and woe betide the child who picked bluebells in the woods for they'd mysteriously vanish into the night! Foxgloves were witches' thimbles – pick a dandelion, make a wish, knock on wood to scare away evil spirits.

Love and passion are ingrained in roses, and *everyone* knows fairies hold their meetings around bluebell rings. Cowslips are so ingrained in my childhood that we really did believe we'd grow taller if we sipped from their sunny cups. I bet you didn't know they sprout in the very spot where St Peter allegedly dropped his keys to the heavenly kingdom (he must have mislaid his keys a great deal, m'thinks).

From the school dormitory we could see Chanctonbury Ring – supposedly the most haunted hill in West Sussex with an ancient stand of circled beech forming a golden halo in summer and wreathed in sinister mist in winter. UFOs had been sighted there! Walk backwards around this famous landmark at midnight and summon the devil – who, it is said, offers a bowl of soup in exchange for children's souls. Elves and fairies are represented as mythical custodians of wildflower landscapes or inhabiting particular plants. Tiny, good folk dwelling in foxgloves, mystical flower sprites and pixies obviously made quite an impression on Shakespeare. He alludes to them in *A Midsummer Night's Dream*:

> And I serve the Fairy Queen, To dew her orbs upon the green. The cowslips tall her pensioners be. In their gold coats spots you see; Those be rubies, fairy favours; In those freckles live their savours. I must go seek some dewdrops here, And hang a pearl in every cowslip's ear.

See? There are fairies at the bottom of the garden – perhaps you can find them more easily now?

JOURNEY'S END:
Seasons in Bloom

It's fitting we find ourselves nearing the end of February as we meander towards journey's end. Hedgerows daubed black with mud, unfulfilled New Year's resolutions squirming in guilty remonstration, bruised clouds and charcoal skies. Then, thrusting into this mournful landscape, the smell of sap rising – the air suddenly filled with unfulfilled promises.

Spring! The sexiest month of the year! The season of renewal and second chances. Bare branches scratch at the sky, leaflets skinny in bud. Brave flowers burst forth. Nodding daffodils sheathed in rough grass, apricot, cream and gold. The best of the daffs are fragrant, 'Cheerfulness' and 'Paper-whites' contrasting with robust 'Churchill', deeply perfumed despite its outward masculinity. Golden celandines, milky snowdrops ice banks and slopes, and dusky Lenten roses. The fruity fragrance of flowering currant coming not from its flowers but crimped foliage. Harbingers of hope and brighter, better days ahead, blazing the trail for the old-fashioneds that come later.

Following hard on the heels of spring bloomers, nobody cares if columbines lack fragrance. From early May, they sashay in flower borders, tossing their bonnets, tittering and batting their eyelashes in bold 'come hither' at bees dowsing among limey boughs. But we must wait a while longer for blossom and roses. Spirited 'Canary Bird', blessed with good-natured abundance, is reason enough to give it room. Apple and cherry blossom flutter from smooth boughs, fragrant woodruff carpets shady glades, colonising by inches rather than yards. Lilacs spritz the air! Spongy green turf and the sweet-breathed sighs of honeysuckle.

Summer drifts in, long and lazy. Puffs of dust scuff under our feet, violet pillows of hebe and thickly tasselled butterfly bushes. Stately spires of delphinium sway with languid grace, lilies waft their ample charms. Brights and bolds mark summer's arrival – day lilies in endless colour parades, with breathtaking blooms lasting only a day. Punches of allium's powder blues, the tulip tree's golden goblets, peeping through leafy green.

Impatient glory bower sheds its fragrant blooms for the cerise-turquoise brooches of autumn. Shimmering grasses stir in breezes, with mounds of stripy cranesbill. Gaudy geums and trembling handkerchiefs of Chinese dogwoods. England – where it's summer in the sunshine and winter in the shade!

Autumn is already scoping, eager to make its return, either scything with cold or with languorous idling as if unaware that winter is hard at its heels. And it brings its own delights. The season of salvias and tuberose, flame-tipped pokers, mercurial Peruvian lilies, angels' fishing rods and crimped begonias. Woodsmoke and diamond-glinted spiders' webs on misty mornings, glossy conkers littering dewed grass. Eye-bright coneflowers, bronzed sneezeweed, tall spires of cerise lobelia and the surprising sunset orange of Kaffir lilies. Indestructible asters and anemones high in the borders gloat that few can match their finery so late in the year. And dahlias – great puffballs of extravagance – with frothing fleabane flowering fearlessly until first frosts.

Winter roars in, blowing icy breath. Shards of lemon in battle-grey skies crack darkening afternoons. It's *that* time of year again when we'll roast chestnuts over open fires and forage for nuts or mistletoe orbed high in the branches. Pearly snowberries squish in hoary grass, the strawberry tree hung with festive baubles. Shrubby honeysuckle, white ballerinas in pristine tutus, and chalices of Christmas roses. Enamelled aconites and Daphne's perfumed waxy stars. Sulphur-yellow candelabras of mahonia tint the air with a delicate hint of lily of the valley tangled with winter clematis, freckled lanterns on bare stems.

A year-long parade of floral fascinations. The cycles of nature – of day and night, the changing seasons. Spring's renewal, summer's abundance, autumn's bounty and winter's rest mirror the stages of human ambition, achievement, growth, preparedness and reflection as we embrace change as a constant force. Before we pack our bags to return home after an incredible journey, remember time with nature is time always well spent – you'll never regret a single moment in its company.

BELIEVE IN TOMORROW: Biomimicry and Advances

In studying the anatomy and flight of birds, Leonardo da Vinci sketched early flying machines that laid the groundwork for the aeroplane. In the same way, modern biologists, environmentalists, scholars, scientists and inventive minds build on the foundations imagined by those before them. Faced with the sobering reality of climate change, we risk overlooking the positive cultural and industrial shifts sparked by its very threat. A challenge that has already given rise to a new generation of thinkers and creators – an era of ingenuity and revival rivalling the Industrial Revolution. Through biomimicry and nature-inspired design, once-unthinkable solutions are becoming real. Sea panels modelled on mangroves are helping rebuild coral reefs and protect coastlines. Fast-fashion brands are turning to algae instead of petroleum-based plastics. Light-capturing satellites, biodegradable plastics, greener chemical processes, wind and hydro power – all point towards a future that harnesses human progress with the planet's needs.

We've experienced a mere snapshot of how plants survive and continue to evolve, and we can learn a lot from them. Create happy, symbiotic relationships when we can, stay versatile and open-minded, adapt when necessary, but know when to dig in and stand firm. Whatever has driven you to discover *The Power of Plants*, it's my unshakeable belief that the world is not just a resource but a living, breathing entity seducing even the most hardened cynic, couch potato or withdrawn child and capable of unleashing potential in every one of us. Nature, with its endless wonders, isn't just a backdrop to human life but an integral part of it, shaping our experiences and inspiring our journey. So easily accessible and commonplace in its abundance, it's easy to overlook our dependence on a world that benignly sustains us. If survival is the essence of all life on Earth, then as we've seen, nature is the difference that makes the impossible possible.

Our oceans teem with life, forests shelter countless creatures, the air we breathe is a daily gift from the plants that purify it. Yet, despite Earth's resilience, its beauty and diversity are fragile. The ecosystems that sustain us require not only our admiration but our active protection and we can't ignore its call for action much longer. In the end, our planet's beauty lies in its balance. And as we look to the future, let's remember that every small action to protect our planet carries immense power. Nature has always given us life and asks for nothing but our respect and care in return, and now it's our chance to give something back. When we choose to care and preserve this incredible, wonderful world of ours, we become part of something far greater: a shared hope for a thriving, vibrant planet that future generations can call home.

AUTHOR'S NOTE

What can I say? Plants make me unaccountably happy. Just when you think you've got their measure, they still have the capacity to delight and surprise. Botanists and scientists – and just ordinary plant lovers like me – have found the plant kingdom to be an endless source of wonder and I hope this book does justice to their astounding capabilities.

Balancing details and ideas across biology, botany, chemistry and science without boring readers to death was always going to be an ambitious undertaking. Even with a respectable amount of horticultural expertise, I've had to delve deep into endless research papers, periodicals and books, not to mention countless online resources, to tell the story of modern plants. The study of plants and wildlife is a vast and intricate field and should you wish to learn more, you'll find my references in the following pages.

Naturally, the huge scope of the plant kingdom means I've only skimmed the surface of their extraordinary beauty, capabilities and resilience. Admittedly, research into plant behaviour is still relatively limited compared to other fields of scientific investigation, and even the most impressive boffins admit that there's still so much we don't know about them – their intelligence, their interactions and adaptations, which contribute to making plants an enduring fascination.

It's encouraging to see that with growing awareness of climate change and the potential potency of plant compounds, experts and global industries are trying to unravel their mysteries for the good of mankind and our planet. The deeper our understanding, the better equipped we are to protect our fragile ecosystems and harness the gifts nature provides. In some small way, this book contributes to supporting their invaluable work and gives us a richer appreciation of our gardens, parks, woodlands and wild spaces.

From the smell of freshly cut grass to the clean scent of wildflowers blooming gaily in a jam jar, we can marvel at their beauty and still ask ourselves, *why do plants do what they do?* Their behaviour and intricate adaptations are still unfolding, and perhaps next time you swat a troublesome wasp, tread on a humble dandelion pushing through a crack in the pavement or ruthlessly mow daisies spotting the lawn, recall the incredible service that plants and nature provide every day. Rooted in resilience, their stories underpin our own.

Each day as I have sat down to write, encountering enchanting plants each with their own personalities and idiosyncrasies – some old friends and many completely new to me – it's been a journey of wonder and discovery. The voyage has been an exciting and humbling one. But you don't need research to simply observe the environment around you, just plain old appreciation and curiosity. Nature watching is addictive, and we should never consider sitting watching the world go by as anything but a necessity. Scuff the grass under your feet, lean back against a tree, close your eyes and breathe in the smells drifting on the breeze. Enjoy being at rest for a few minutes surrounded by nature.

My thanks to my editor Claire for sharing my vision and her unflagging enthusiasm for my plant stories, and to the many authors, researchers, botanists and scientists whose research I've consulted to make this book possible. And, of course, my genuine thanks to you for picking up *The Power of Plants* and reading. It makes all the months spent writing and researching this happy compendium worthwhile. If you'd like to know more about what I get up to, skip over to www.lucysummers.co, where you can keep up with new books and other events you may enjoy.

COPYRIGHTS

My thanks to the following individuals and organisations who have so generously allowed the use of their clients' work in this book:

The Society of Authors, as the literary representative of the estate of Rose Fyleman.

Amanda Wallace at Thrive UK, together with Abigail Garvey, Cynthia Coker and Joseph Raphael for sharing their experiences.

GLOSSARY

ADHD (Attention-Deficit/Hyperactivity Disorder): A condition where the brain works differently to most people.
Ambergris: A substance formed in sperm whale intestines.
Ambrein: An odourless alcohol extract.
Angiosperms: A group of plants that have flowers and seeds.
Anthocyanin: Water-soluble scarlet, magenta, purple and blue pigments that colour the fruit and flowers of many plants and also provide the red colours of many autumn leaves.
Apomorphine: An alkaloid found in blue water lilies.
BDNF (brain-derived neurotrophic factor): A protein that supports the survival and growth of nerve cells (neurons) in the brain and spinal cord.
Bipedalism: The ability to walk on two legs.
Blossom end rot: A calcium deficiency found in vegetables such as aubergines, peppers and more commonly, tomatoes.
CAM (Crassulacean Acid Metabolism): A device to minimise photorespiration.
Cambium: Layers of cells in the roots and stems that produce new tissue.
Carbon dioxide (CO2): A heat-trapping gas, also called greenhouse gas.
Chlorophyll: A green light-absorbing pigment.
Dicot: A flowering plant with two seed leaves (cotyledons) emerging from its embryo.
Diosgenin: A plant-based steroid.
Elastiome: A protein-rich appendage found in seeds.
Euglossine bee: Orchid bee.
Epiphyte: A plant that grows on another plant, often one that is not parasitic.
Fallopian tubes: A pair of tubes in a female mammal through which eggs travel from the ovaries to the uterus.
Fragmentation: A form of asexual reproduction wherein a parent organism breaks into fragments, each capable of growing independently into a new organism.

Geotropism: The directional growth of plants in response to gravity.
Gravitropism: The movement or growth of a plant in response to gravity.
Gymnosperma: A group of plants that have seeds that are unprotected by an ovary or fruit, including conifers and cycads.
Haustorium: A slender projection from the root of a parasitic plant.
Hemiparasitic: Plants that rely on a host for water and mineral nutrients, taking some (but not all) of their nourishment requirements from the host plant.
Holoparasite: A plant that is incapable of photosynthesis and obtains all its nutrients and water from a host plant.
Linnaeus, Carl: An eighteenth-century Swedish botanist.
NHS (National Health Service): Publicly funded healthcare system of the UK.
Malic acid: Found in many fruits, such as apple, and contributes to the sour taste of green apples.
Monocot: A flowering plant with one seed leaf (cotyledon) emerging from its embryo.
Mycobacterium vaccae: A soil-derived bacterium with stress-resilient properties.
NASA (National Aeronautics and Space Administration): A United States government agency responsible for the nation's civilian space programme, aeronautics and space research.
Negative gravitropism: Negative gravity against a gravitational upward force; when the stem of a plant grows *against* the force of gravity (i.e. upwards).
Obligate parasite: An organism that cannot survive independently of a host species.
Phloem: Tissues in plants that conduct the food made in the leaves to all the other parts of the plant.
Phytochrome: Light-sensitive photoreceptor found in plants, bacteria and fungi.
Pistil: The female reproductive organs of a flower, comprising stigma, style and ovary.
Positive gravitropism: When the root pf a plant grows in the direction of the force of gravity (i.e. downwards).
Powdery mildew: A fungal disease affecting a wide range of plants.
Prochlorococcus: A small, unicellular marine cyanobacterium, which is considered to be one of the most abundant photosynthetic organisms on earth.
Propanethial-S-oxide: A volatile gas that triggers tears, for example when an onion is cut.
Prothallia: The small, green, heart-shaped structure (gametophyte) of a fern that produces both male and female sex cells (gametes).
PTSD (Post-Traumatic Stress Disorder): A condition caused by a traumatic event.

Rhizomorph: A long strand sent out by some fungi to colonise new space and absorb nutrients.
Scarification: A process whereby seed dormancy is broken to promote germination via methods such as nicking, breaking, softening or otherwise weakening of the seed coating.
Sesquiterpene lactones: Plant-derived compounds used in medicine.
Spiny pocket mouse: A rodent from the *Heteromys* family.
Spores: Cells produced by seedless plants with the ability to grow into a new plant or organism.
Stamen: The male reproductive organ of a flower.
Stigma: The female sex organs of a plant.
Stratification: The process in which a seed is exposed to moist, cold or warm conditions in order to break dormancy.
Taxol: A compound found in the bark of the Pacific yew which is found to inhibit certain cancers.
Teosinte: Wild grasses closely related to corn, found in Mexico and Central America.
Tulip fire: Fungal disease of tulips caused by *Botrytis tulipae*, which causes brown spots and twisted, withered or distorted leaves.
WHO (World Health Organization): A specialised agency of the United Nations to coordinate public health initiatives worldwide.
Xylem: Plant vascular tissue that conveys water and dissolved minerals from the roots to the rest of the plant and also provides physical support.

GLOSSARY OF PLANT NAMES

African baobab: *Adansonia digitata*, a distinctive and iconic tree native to Arabia, Madagascar, mainland Africa and Australia.
Air plant: *Tillandsia* spp., a genus of around 650 species of evergreen, perennial flowering plants in the family Bromeliaceae from Mexico and South America.
Alligator plant: *Kalanchoe daigremontiana*, a succulent plant native to Madagascar and Mexico that can propagate from plantlets on its leaves.
Angels Trumpets: *Brugmansia*, a genus of flowering plants, woody trees or shrubs in the nightshade family (Solanaceae) from South America.
Argan tree: *Argania spinosa*, a thorny tree from south-western Morocco and Algeria.
Australian Christmas tree: *Nuytsia floribunda*, an evergreen hemiparasitic tree native to Western Australia.
Azalea: Evergreen or deciduous shrubs related to rhododendrons from Asia.
Bindweed: A twining or creeping herbaceous perennial plant with trumpet-shaped flowers, native to Europe and Asia. Some species are invasive.
Bird of Paradise: *Strelitzia reginae*, a flowering perennial, native to the Cape Provinces and KwaZulu-Natal in South Africa.
Borneo orchid: *Orchidantha inouei*, a poorly known flowering perennial found across southern China to Borneo.
Bull bay: *Magnolia grandiflora*, an evergreen tree, native to the south-eastern United States.
Bulrush or reed mace: *Typha latifolia*, a herbaceous perennial wetland plant found throughout most of Europe, Eurasia and North America, and more locally in Africa and South America.
Catmint: *Nepeta*, a group of flowering perennials native to Europe, Asia and Africa. Some species are known as catnip or catmint because they affect cats' behaviour with nepetalactone.

Chinese camellia: *Camellia sinensis*, an evergreen flowering shrub or small tree native to southern parts of China, Japan and Korea from which tea has been produced for thousands of years.
Chinese wisteria: *Wisteria sinensis,* a deciduous vine native to China with fragrant spring flowers.
Columbine: *Aquilegia* spp., commonly known as columbines, they are flowering perennial plants found across the Northern Hemisphere.
Colombian coca: *Erythroxylum novogranatense*, an evergreen shrub native to Colombia and Venezuela.
Common bladderwort: *Utricularia vulgarisa,* aquatic species of bladderwort found in Asia and Europe.
Common tobacco: *Nicotiana tabacum*, annually grown herbaceous plant native to South America and widely grown for commercial tobacco production.
Corpse flower: *Rafflesia arnoldii*, a flowering, parasitic plant native to the rainforests of Sumatra and Borneo.
Deadly nightshade: *Atropa bella-donna*, a toxic perennial plant native to Europe and West Asia.
Desert moss: *Tortula muralis,* a species of moss found worldwide.
Dittany: *Dictamnus albus*, gas plant or burning bush, a herbaceous perennial plant native to warm woodland habitats across southern Europe, North Africa and Asia.
Dodder: *Cuscuta,* yellow, orange or red parasitic plants found throughout temperate and tropical regions of the world.
Dragon blood tree: *Dracaena cinnabari*, a unique, native tree almost extinct everywhere except in small habitats of Socotra Archipelago (Yemen), a UNESCO World Heritage Site.
Dypsis minuta: No common name, a miniature palm from the rainforests of Madagascar.
Elephant's ears: *Alocasi*, a broad-leaved, flowering perennial native to tropical and subtropical Asia and eastern Australia.
Fever tree: *Cinchona pubescens*, a deciduous, medicinal tree native to Central and South America.
Firethorn: *Pyracantha* spp., thorny, evergreen shrubs extending from south-west Europe and Southeast Asia.
Fumewort: *Corydalis hemidicentra*, an alpine plant found across south-east Tibet to China.
Garden cress: *Lepidium sativum*, a fast-growing, edible herb native to West Asia.
Ghost flower: *Monotropa uniflora*, a holoparasitic, flowering perennial native to temperate regions of European Russia, Asia, North America and northern South America.

Golden perennial hop: *Humulus lupulus* 'Aureus', a twining, deciduous vine native to Europe, West Asia and North America.
Goldenrod: *Solidago canadensis*, an invasive flowering perennial, native to North America.
Green rose: *Rosa* 'Viridiflora', a twiggy, almost thornless rose native to China.
Himalayan blue poppy: *Meconopsis baileyi*, a clump-forming, flowering perennial from the Himalayas.
Hornwort: *Ceratophyllum demersum*, a submerged aquatic plant native to all continents except Antarctica.
Hurricane palm: *Dictyosperma album*, also called the princess palm, a critically endangered palm, native to the Mascarene Islands.
Japanese wisteria: *Wisteria floribunda*, a twining, woody, deciduous flowering plant native to Japan.
Little coca: *Erythroxylum coca*, a tropical evergreen shrub native to western South America.
Liverwort: Species of small, non-vascular, spore-producing plants.
Maidenhair fern: *Adiantum raddianum*, a perennial fern native to Trinidad, Tobago, Mexico and South America
Maltese fungus: *Cynomorium coccineum*, a parasitic flowering perennial native to Lanzarote, the Mediterranean, Mauritania and Mongolia.
Marijuana: *Cannabis sativa*, an annual flowering plant native to East Asia.
Mexican chocho palm: *Astrocaryum mexicanum*, a long-lived palm native to Mexico and Central America.
Mexican jumping bean shrub: *Sebastiana pavoniana*, a shrub or small tree native to tropical regions of South America.
Miracle tree: *Leucaena leucocephala*, a shrub or tree native to Mexico and Central America.
Mistletoe: *Viscum album*, an evergreen, hemi-parasitic plant native to Europe, Lebanon, Nepal and north-east Algeria.
Morning glory: *Ipomoea purpurea*, a perennial, twining climbing plant native to Mexico and Central America.
Portuguese laurel: *Prunus lusitanica*, an evergreen shrub or small tree, native to Portugal and western Spain.
Quindío wax palm: *Ceroxylon quindiuense*, a palm native to Colombia and Peru.
Red oak: *Quercus rubra*, a deciduous tree native to North America, eastern and central United States, south-east and south-central Canada.
Resurrection plant: *Selaginella lepidophylla*, a desert plant native to the Chihuahuan Desert of the United States.
Rhododendron: A family of evergreen shrubs originating from Asia.
Rock cress: *Arabidopsis thaliana*, a small annual plant native to Eurasia.

Seagrass: *Zostera*, eelgrass, one of world's most widespread marine plant species, predominantly found along the temperate and cold coasts of the Northern Hemisphere including the Atlantic Ocean, Pacific Ocean, Baltic Sea and Mediterranean Sea.
Silk tassel bush: *Garrya eliptica*, a fast-growing evergreen shrub native to California and southern Oregon.
Silverbush: *Convolvulus cneorum*, a small evergreen shrub native to coastal areas of Spain, Italy, Croatia and Albania.
Slipper orchid: *Cypripedium subtropicum*, an endangered orchid native to south-east Tibet, China (specifically south Yunnan) and North Vietnam.
Snot goblin: *Persoonia pinifolia*, a large evergreen flowering shrub native to Australia.
Spanish flag: *Lantana camara*, an evergreen flowering shrub native to Central and South America.
St Helen's ebony: *Trochetiopsis ebenus*, a flowering shrub native to Saint Helena in the South Atlantic Ocean.
Steppe screw moss: *Syntrichia caninervis*, a desert moss found in temperate regions worldwide.
Stinkhorn mushroom: *Phallus impudicus*, a phallic-shaped fungus found in temperate regions of Europe and North America.
Strychnine tree: *Strychnos nux-vomica*, an evergreen or deciduous tree with toxic fruit, native to India and Southeast Asia.
Sundew: *Drosera rotundifolia*, a carnivorous, flowering plant growing in bogs, marshes and fens, native to the UK and found in much of Europe, North Asia and North America.
Sunflower: *Helianthus annuus*, a flowering annual, widely grown as an oilseed crop, native to North America and cultivated throughout the world.
Sweetcorn: *Zea mays*, an annual flowering plant native to South Mexico.
Sweet flowering tobacco: *Nicotiana alata*, a fragrant perennial native to South America.
Swiss cheese plant: *Monstera deliciosa*, a tropical flowering perennial native to the rainforests of Central America, from southern Mexico to Panama. Commonly grown as a houseplant.
Teosinte: Wild grasses closely related to corn, found in Mexico and Central America.
Toothwort: *Lathraea clandestine*, a parasitic flowering perennial native to western France.
Traveller's palm: *Ravenala madagascariensis*, a flowering palm native to Madagascar.
Tree of heaven: *Ailanthus altissima*, a spreading, deciduous tree native to China, North Vietnam and Taiwan.

Viper's bugloss: *Echium vulgare*, a flowering perennial native to the UK.
Voodoo or stink lily: *Dracunculus vulgaris*, a flowering perennial native to Greece, Crete, the Aegean Islands and Turkey.
W. 'General Sherman': *Wellingtonia sequoia*, Redwood, an evergreen tree native to the Sierra Nevada mountains of California.
Walking palm: *Socratea exorrhiza*, a palm native to the rainforests of tropical Central and South America.
Watermeal: *Wolffia* spp., aquatic plants native to Asia.
Winter-flowering honeysuckle: *Lonicera fragrantissima*, a flowering shrub native to Asia.
Woodruff: *Galium odoratum*, a woodland, flowering perennial native to Europe, North Africa and parts of Asia.
Yellow rattle: *Rhinanthus minor*, an annual wildflower native to Europe and West Asia.

REFERENCES

BOOKS

Abe, Naoke, *'Cherry' Ingram: The Englishman Who Saved Japan's Blossoms* (Vintage, 2020).

Beeckman, Tom, & Amram Eshel, *Plant Roots: The Hidden Half* (Fifth edition, CRC Press, 2024).

D'amato, Peter, *The Savage Garden: Revised – Cultivating Carnivorous Plants* (Ten Speed Press, 2013).

Darwin, Charles, *On the Origin of the Species: The Science Classic*, with introduction by John van Whye (Capstone Classics, 2019).

Darwin, Charles, & Sir Francis Darwin, *The Power of Movement in Plants: Phototropism in Plants* (annotated), (1880: BoD, 2022 edition).

Darwin, Sir Francis (ed.), *Life and Letters of Charles Darwin: Volume III* (John Murray, 1901).

Dawson, Peter, *A Handbook for Horticultural Students* (Peter Dawson, 2023).

Emerson, Waldo Ralph, *Essays: Second Series* (James Munroe & Company, 1844).

Forbes, Peter, *Dazzled and Deceived: Mimicry and Camouflage* (Yale University Press, 2011).

Hanson, Thor, *Buzz: The Nature and Necessity of Bees* (Icon Books, 2019).

Jones, Richard, *Ants: The Ultimate Social Insects* (British Wildlife Collection) (Bloomsbury Wildlife, 2022).

King, F.H., *Farmers of Forty Centuries: Organic Farming in China, Korea, and Japan* (Dover Publications, Inc., 2004).

Moore, Thomas, *Soul Mates* (Harper Perennial, 2016).

Monbiot, George, *Feral: Rewilding the Land, Sea and Human Life* (Penguin, 2014).

Monbiot, George, *Regenesis: Feeding the World without Devouring the Planet* (Penguin, 2023).

Paige, Russell, *The Education of a Gardener* (Vintage Classics, 2023).

Pliny, the Elder, *The Natural History of Pliny, Vol. II* (Henry G. Bohn, 1855).

Schlanger, Zoë, *The Light Eaters: The New Science of Plant Intelligence* (Fourth Estate, 2024).
Sheldrake, Merlin, *Entangled Life: How Fungi Make Our Worlds, Change Our Minds and Shape Our Futures* (Bodley Head, 2023).
Sverdrup-Thygeson, Anne, *Extraordinary Insects* (Mudlark, 2019).
Terrill, Andrew, *The Earth Beneath My Feet: A 7,000-Mile Walk of Discovery into the Heart of Wild Nature* (Enchanted Rock Press, 2022).
Wilson, Edward O., *Tales from the Ant World* (Liveright, 2020).
Wohlleben, Peter, *The Hidden Life of Trees: What They Feel, How They Communicate* (William Collins, 2017).

ARTICLES

Banga, O., 'Origin of the European Cultivated Carrot', *Euphytica,* Vol. 6 (Institute of Horticultural Breeding, 1956), pp. 55–57.
Bergman, Melanie, Lars Gutow & Michael Klarges (eds), *Marine Anthropogenic Litter* (eBook, Springer, 2015) pp. 5–159 (accessed 1 January 2015).
Chehab, E. Wassim, Elizabeth Eich & Janet Braam, 'Thigmomorphogenesis: A Complex Plant Response to Mechano-Stimulation', *Journal of Experimental Botany*, Vol. 60, Issue 1 (January 2009) pp. 43–56.
Cordes, Eugene H., 'The Discovery of Taxol: Wall, Horwitz, Holton', *Hallelujah Moments: Tales of Drug Discovery*, 2nd edition (online edn, Oxford Academic, 2021), pp. 103–20.
Dani, John A., et al., 'Historical and Current Perspective on Tobacco Use and Nicotine Addiction', *Trends in Neurosciences*, Vol. 34, Issue 7, pp. 383–92.
Daniel, Robert, 'Plant Bioacoustics: The Sound Expression of Stress', *Cell*, Vol. 186, No. 7 (30 March 2023).
Desborough, M.J.R., & D.M. Keeling, 'The Aspirin Story: From Willow to Wonder Drug', *British Journal of Haematology* (Epub January 2017), pp. 2–5.
Friedman, J., & M.J. Rubin, 'All in Good Time: Understanding Annual and Perennial Strategies in Plants', *American Journal of Botany*, Vol. 102 (2015) pp. 497–99.
Glasgow Corporation, 'Baths and Wash Houses Annual Reports, 1900 to 1961'.
Jiang, Hong, Ji-Jun Kong, Hsin-Chun Chen, et al., '*Cypripedium subtropicum* (*Orchidaceae*) employs aphid colony mimicry to attract hoverfly (*Syrphidae*) pollinators', *New Phytologist,* Vol. 227, Issue 4 (2020), pp. 1213–21, doi: 10.1111/nph.16623.
Livesey, P., 'Milestones in the History of Concrete Technology: The Origins of Portland Cement', *Institute of Concrete Technology Yearbook 2002–03*, pp. 13–21.

Locock, Martin, 'The Development of the Building Trades in the West Midlands, 1400–1850', *Construction History*, Vol. 8 (1992) pp. 3–20.

Nails, Debra, & S. Sara Monoson, 'Socrates' in Edward N. Zalta & Uri Nodelman (eds), *The Stanford Encyclopedia of Philosophy* (Summer 2022 edition).

Ramirez, J.C., Hugge, MD Christopher, 'Aspirin: The Medicine That Would Not Die', *The Journal of Pediatrics* (January 2022), p. 65.

Schöner, Michael G., Ralph Simon & Caroline R. Schöner, 'Acoustic Communication in Plant–Animal Interactions', *Current Opinion in Plant Biology*, Vol. 32 (August 2016), pp. 88–95.

Thiery, M., 'Gabriele Fallopio (1523–1562) and the Fallopian Tube', *Journal of Gynecological Surgery*, Vol. 6, pp. 93–95 (2009), https://doi.org/10.1007/s10397-008-0453-3

Wotton, Karl R., et al., 'Mass Seasonal Migrations of Hoverflies Provide Extensive Pollination and Crop Protection Services', *Current Biology*, Vol. 29, Issue 13, pp. 1–6.

WWF (World Wide Fund for Nature), *Living Planet Report 2020: Bending the Curve of Biodiversity Loss*, edited by R.E.A. Almond, M. Grooten & T. Petersen (Gland, Switzerland: WWF, 2020), pp. 1–21.

Yager, Karl Graeme, *Batesian Mimicry in Plants* (University of Wellington, Thesis, 2018), pp. 3–129: https://openaccess.wgtn.ac.nz/articles/thesis/Batesian_mimicry_in_plants_The_significance_of_shared_leaf_shape_between_Alseuosmia_pusilla_and_Pseudowintera_colorata/17072279?file=31565477

SCIENTIFIC/ACADEMIC PAPERS

Bratman, Gregory N., et al., 'Nature and Human Well-Being: The Olfactory Pathway', *Science Advances*, Vol. 10, Issue 20 (May 2024): https://doi.org/10.1126/sciadv.adn3028

Darwin, Charles, 'Unpublished Correspondence of Charles Darwin', reprinted from *Natural History*, Vol. XXIII, No. 26 (1923), pp. 589–96.

Elliott, Daniela Dutra, & Paula Mejia Velasquez, *Botany in Hawai'i* (Leeward Community College, https://bio.libretexts.org).

Kollasch, A.M., A.R. Abdul-Kafi, M.J.A. Body., et al., 'Leaf Vibrations Produced by Chewing Provide a Consistent Acoustic Target for Plant Recognition of Herbivores', *Oecologia* 194(1–2):1–13 (October 2020) doi: 10.1007/s00442-020-04672-2

Lee, J.S., H.A. Oh, J.Y. Kwon, et al., 'The Effects of *Cynomorium songaricum* on the Reproductive Activity in Male Golden Hamsters', *Development & Reproduction* 17(1):37–43 (March 2013) doi: 10.12717/DR.2013.17.1.037

McElrone, A.J., B. Choat, G.A. Gambetta & C.R. Brodersen, 'Water Uptake and Transport in Vascular Plants', *Nature Education Knowledge* 4(5):6 (2013).

Poulin, Robert, 'The Many Roads to Parasitism: A Tale of Convergence', *Advances in Parasitology*, Vol. 74 (2011), pp. 1–40.

WEBSITES

American Chemical Society, 'Joseph Priestley and the Discovery of Oxygen'. Available at: www.acs.org/education/whatischemistry/landmarks/josephpriestleyoxygen.html (accessed 12 November 2025).

American Museum of Natural History, 'Dispatches from the Deep: Light and Dark in the Sea'. Available at: www.amnh.org/learn-teach/curriculum-collections/deep-sea-vents/light-and-dark-in-the-sea (accessed 27 April 2025).

American Psychological Association, 'Nurtured by Nature' Available at: www.apa.org/monitor/2020/04/nurtured-nature (accessed 27 May 2024).

Association of Anaesthetists, 'A Brief History of Morphine'. Available at: https://anaesthetists.org/Home/Heritage-centre/Learning/Heritage-Centre-blog-celebrating-75-years-of-Anaesthesia/Fatal-Attraction-A-Brief-History-of-Morphine-a-blog-post-from-the-Heritage-Centre (accessed 15 June 2024).

Atomic Energy Foundation, 'Otto Hahn'. Available at: https://ahf.nuclearmuseum.org/ahf/profile/otto-hahn/ (accessed 14 December 2024).

Baobab Foundation, 'The real truth about water in baobab trees'. Available at https://baobabfoundation.co.za/natures-water-tank (accessed 11 January 2024).

British Ecological Society, 'Ants are the major agents of resource removal from tropical rainforests'. Available at: https://besjournals.onlinelibrary.wiley.com/doi/full/10.1111/1365-2656.12728 (accessed 8 October 2025).

Cambridge University Botanic Garden, 'Sensitive Plant, Mimosa pudica)' Available at: www.botanic.cam.ac.uk/learning/trails/plant-rhythms/sensitive-plant (accessed 11 January 2025).

MIT Technological Review, 'Planting more trees could suck up a huge share of carbon emissions'. Available at: https://climate.mit.edu/ask-mit/how-many-new-trees-would-we-need-offset-our-carbon- (accessed 10 April 2025).

Department of Biology, University of Oxford, 'Unearthing the social life of ants'. Available at: www.biology.ox.ac.uk/article/unearthing-the-social-life-of-ants (accessed 29 October 2025).

Drug Enforcement Administration Museum, 'Opium poppy'. Available at: https://museum.dea.gov/exhibits/

online-exhibits/cannabis-coca-and-poppy-natures-addictive-plants/opium-poppy#:~:text=The%20earliest%20reference%20to%20opium,it%20on%20to%20the%20Egyptians (accessed 10 January 2024).

Dung Beetle Solutions International, 'Dung Beetle Solutions International'. Available at: www.dungbeetlesolutions.com.au/about-dung-beetles/bush-flies-and-dung-beetles (accessed 13 June 2024).

Entomology Today, 'Ants can lift up to 5000 times their own bodyweight'. Available at: https://entomologytoday.org/2014/02/11/ants-can-lift-up-to-5000-times-their-own-body-weight-new-study-suggests (accessed 3 February 2025).

European Commission, 'The Hidden Part of the Climate Cycle'. https://climate.ec.europa.eu/system/files/2016-11/soil_and_climate_en.pdf (accessed 2 January 2025).

Florida Museum 'The secret lives of geophytes'. Available at: www.floridamuseum.ufl.edu/science/the-secret-lives-of-geophytes (accessed 7 June 2024).

Government of Western Australia, 'Sea anemones'. Available at: www.dbca.wa.gov.au/wildlife-and-ecosystems/marine/marine-parks/fun-facts/sea-anemones (accessed 5 January 2025).

Henry Shaw Cactus & Succulent Society, 'C3, C4, CAM Photosynthesis'. Available at: https://hscactus.org/resources/digest/plant-info/c3-c4-cam (accessed 27 April 2024).

History Today, 'The English Physician'. Available at: www.historytoday.com/history-matters/english-physician (accessed 5 October 2024).

Maryland Agronomy News, 'Wasps, surprisingly cool pollinators'. Available at: https://blog.umd.edu/agronomynews/2020/08/31/wasps-surprisingly-cool-pollinators (accessed 5 September 2025).

National Archives, 'Homestead Act (1862)'. Available at: www.archives.gov/milestone-documents/homestead-act (accessed March 8 2024).

National Geographic, 'Isaac Newton: Who was he?'. Available at: https://education.nationalgeographic.org/resource/isaac-newton-who-he-was-why-apples-are-falling/4th-grade (accessed 17 October 2024).

Natural History Museum, 'How an Asteroid Caused the Extinction of the Dinosaurs'. Available at: www.nhm.ac.uk/discover/how-an-asteroid-caused-extinction-of-dinosaurs.html (accessed 5 September 2024).

Natural History Museum, 'Tumbleweeds: The fastest plant invasion in the USA's history'. www.nhm.ac.uk/discover/tumbleweeds-fastest-plant-invasion-in-usa-history.html (accessed 6 March 2025).

Oxford University Academic, *Journal of Experimental Botany*, 'Thigmomorphogenesis: a complex plant response to mechano-stimulation'.

https://academic.oup.com/jxb/article/60/1/43/572567?login=false (accessed 9 October 2024).

Penn Museum, 'The Embattled Driver in Ancient Rome'. Available at: www.penn.museum/documents/publications/expedition/2-3/The%20Embattled.pdf (accessed 19 January 2024).

Phys.org, 'A Matter of Gravity – Understanding How Plants Grow in Space'. Available at: https://phys.org/news/2018-07-gravityunderstanding-space.html (accessed 11 December 2024).

Project Seagrass, 'Fauna Returns Rapidly in Planted Seagrass Meadows'. Available at: www.projectseagrass.org/research/fauna-return-rapidly-in-planted-seagrass-meadows-study-shows (accessed 27 April 2025).

Royal Botanical Gardens Kew, 'Evolution and diversification of the palms and their seed traits'. Available at: www.kew.org/science/our-science/projects/evolution-and-diversification-palms-seed-traits (accessed 8 April 2024).

Science History Institute, Museum and Library, 'Felix Hoffman'. Available at: www.sciencehistory.org/education/scientific-biographies/felix-hoffmann/ (accessed 8 April 2024).

Science org, 'Shhh, the Ants Are Talking'. Available at: www.science.org/content/article/shhh-ants-are-talking (accessed 11 May 2024).

Scientific American, 'Stressed Plants 'Cry' – and Some Animals Can Probably Hear Them'. Available at: www.scientificamerican.com/article/stressed-plants-cry-and-some-animals-can-probably-hear-them (accessed 6 February 2024).

Shark Trust, 'Shark Senses'. Available at: www.sharktrust.org/shark-senses (accessed 16 April 2025).

Smithsonian Magazine, 'This Enzyme Is Why Onions Make You Cry'. Available at: www.smithsonianmag.com/smart-news/researchers-finally-figure-out-onion-enzyme-makes-you-cry-180964325/#:~:text=As%20Ashton%20Yoon%20reports%20for,acids%20to%20create%20sulfenic%20acid (accessed 15 September 2025).

Smithsonian Natural Museum of History, 'Sea Turtles'. Available at: https://ocean.si.edu/ocean-life/reptiles/sea-turtles (accessed 27 April 2025).

Statista, 'Time spent consuming media per week among children in the United Kingdom (UK) in 2018'. Available at: www.statista.com/statistics/397851/hours-of-media-consumption-by-children-by-media-uk (accessed 16 June 2024).

The Chelsea Physic Garden, 'The species we mourn and the ones we blame'. Available at: www.chelseaphysicgarden.co.uk/the-species-we-mourn-and-the-ones-we-blame (accessed 1 October 2024).

The Huntingdon, 'Darwin's Garden: An Evolutionary Adventure'. Available at: https://huntington.org/exhibition/darwins-garden-evolutionary-adventure (accessed 17 October 2024).

UK Government, 'England's Largest Ever Seagrass Planting Hits New Milestone'. Available at: www.gov.uk/government/news/englands-largest-ever-seagrass-planting-hits-new-milestone--2 (accessed 11 November 2024).

University of Leeds, 'How Plants Conquered the Land'. Available at: https://biologicalsciences.leeds.ac.uk/biological-sciences/news/article/232/how-plants-conquered-the-land#:~:text=It's%20difficult%20to%20imagine,a%20hormone%20that%20prevents%20dehydration (accessed 5 September 2024).

University of Leeds, 'Land-sea "tag-team" devastated ocean life millions of years ago'. Available at: www.leeds.ac.uk/research-32/news/article/5639/land-sea-tag-team-devastated-ocean-life-millions-of-years-ago (accessed 9 November 2024).

University of Oxford, 'Tulipmania: A Garden Historian's Perspective'. Available at: www.history.ox.ac.uk/tulipmania-garden-historians-perspective#:~:text=The%20events%20of%20Tulpenwoerde%20(tulip,the%20irrationality%20of%20economic%20man (accessed 12 June 2024).

University of Oxford, 'Courting bugs attract mates using an elastic 'snapping organ' for vibrational communication'. Available at: www.ox.ac.uk/news/2019-03-13-courting-bugs-attract-mates-using-elastic-snapping-organ-vibrational-communication#:~:text=Planthopper%20bugs%20may%20be%20small,rhythmic%20motions%20of%20their%20abdomen (accessed 4 November 2024).

University of Oxford, 'Plants, sex and Darwin'. Available at: www.ox.ac.uk/news/science-blog/plants-sex-darwin (accessed 1 September 2024).

University of Oxford, 'Plants 400', 'Ficus benghalensis' (Moraceae). Available at: https://herbaria.plants.ox.ac.uk/bol/plants400/Profiles/EF/Ficusb#:~:text=Banyans%20are%20pollinated%20by%20the,are%20digested%20by%20the%20fig (accessed 17 October 2024).

University of Washington, 'Plants and spontaneous combustion'. Available at: https://depts.washington.edu/hortlib/pal/pal-question-1013 (accessed 4 June 2025).

University of Wisconsin-Madison, 'Common Purslane, Portulaca oleracea'. Available at: https://hort.extension.wisc.edu/articles/common-purslane-portulaca-oleracea (accessed 3 March 2025).

University of York, School of Physics, Engineering & Technology: Isaac Newton's apple tree'. Available at: www.york.ac.uk/physics-engineering-technology/about/newtons-apple-tree/#:~:text=Newton's%20 discovery&text=The%20first%20written%20account%20appears,the%20 late%20summer%20of%201666 (accessed 7 February 2024).

Wine & Spirit Education Trust, 'What is Poitín? A quick guide to the traditional Irish spirit'. Available at: www.wsetglobal.com/knowledge-centre/blog/2023/june/15/what-is-poitin-a-quick-guide-to-the-traditional-irish-spirit (accessed 9 June 2025).

Wine & Spirit Education Trust, 'Feet first – Why some winemakers still crush grapes by foot). www.wsetglobal.com/knowledge-centre/blog/2024/feet-first-why-some-winemakers-still-crush-grapes-by-foot (accessed 23 October).

Woodland Trust, 'Why we need trees'. Available at: www.woodlandtrust.org.uk/trees-woods-and-wildlife/british-trees/benefits (accessed 13 March 2025).

Wyss Institute, Harvard University, 'Bioplastics: Environmentally-friendly plastics that biodegrade and require less energy to make'. Available at: https://wyss.harvard.edu/technology/bioplastic (accessed 10 June 2025).

ABOUT THE AUTHOR

Lucy Summers is an RHS qualified horticulturalist, Chelsea gold medallist and award-winning garden journalist. She has presented gardening and consumer programmes for Channel 4, ITV, BBC and radio, and gives regular talks to companies and hobby clubs, as well as writing for print and online media.